construction geometry

Second Edition

construction geometry

Brian Walmsley, C.E.T. C.&G.

Centennial College Press

Published by Centennial College Press
951 Carlaw Avenue
Toronto, ontario
M4K 3M2

All rights reserved. No part of this book may be reproduced or transmitted in any form or by any means electronic or mechanical, including photocopying, recording or any information storage and retrieval system without permission in writing from the author.

Copyright © 1981 Brian Walmsley
Revised Edition 1999

ISBN 0-919852-19-X

Also by Brian Walmsley
The Technical Carpenter
ISBN 0-919852-28-9

Printed in Canada
By Maracle Press Limited

To Brenda

And to apprenticeship, a training concept
that is very meaningful to me.

I am grateful for the constructive criticism of
my colleagues, in particular,
Mr. A. Adams and Mr. J. McIntyre.

Contents

Objectives

The student will solve a set number of problems related to each completed unit within a given tolerance.

Operations

The student should follow the step by step instructions related to the illustrations to lay out every problem. It will be noted that the visual picture condenses the need for lengthy directions.

At the completion of each unit, the student will attempt a test to measure achievement.

Equipment

Though draftsmanship is not stressed in its entirety here, accuracy is essential if geometry is going to work.

It is recommended that to develop a sense of layout, proportion and neatness, a student must acquire and use a T-square, large span compass, 45° set square, 30°—60° set square, architect's scale rule and several pencils. Hard pencils, no. 6H or no. 4H, are necessary for light construction lines, and softer pencils, no. 2H, for more solid lines. A sharp pencil is necessary at all times.

Preface

In the following units, dimensions are suggested purely as a guide to the exercise and may be varied according to preference. Though dimensions are always critical in the end result, emphasis here is placed on method. It should also be appreciated that there is often more than one approach to the solution of a problem and all methods cannot be shown. Metric measurements are also suggested not as a comparison, but to provide an opportunity to practise with S.I. (Systems International) units.

Conversion charts at back.

Introduction

Though the works included here are applicable to many industrial fields, this book is intended as an aid to students of the building industry. With careful study, the units will help to improve the mind and develop an understanding of geometrical applications.

Layout situations and often mathematical problems, can readily be solved by simple drawing if one has learned some basic geometry.

A knowledge of drawing instruments is an asset, but not necessary. It will be seen that most of the following problems can be solved with the aid of compass and ruler, or in larger form, a chalk line radius rod and measuring tape.

Starting with simple problems each section moves slowly to the more complex. This method of a practically applied geometry lends itself directly to people in carpentry, millwork, building, drafting and almost anyone concerned with the building industry.

Definition of Terms and Symbols

Point	A point has position, but no dimension.
Line	A line has no set thickness, but has length; it can be curved or straight. A straight line is the shortest distance between two points.
Surface	A surface may have dimension in several directions, but has no thickness. A flat surface is often referred to as a plane; if it lies flat and level it is a horizontal plane; if the surface is up and down it is a vertical plane; if it lies at an angle it is an inclined plane.
Solid	A solid has three dimensions; length, width and thickness.
Bisect	'Bi' means two, 'sect' means divide or separate, thus, to bisect means to divide into two equal parts.
Trisect	'Tri' means three, so to trisect means to divide into three equal parts.
Notation	To use notation means to letter or number the diagrams at points, intersections, etc. Lettering and numbering should be done without faulter as the drawing is developed — this eliminates many possible errors.
Right Angle	A 90° angle — a square corner.
Acute Angle	Any angle which is less than 90°.
Obtuse Angle	Any angle which is greater than 90°, but less than 180°.
Complimentary Angles	Two angles whose sum is equal to 90°, each is complimentary to the other.
Supplementary Angles	Two angles whose sum is equal to 180°, each is supplementary to the other.
Straight Angle	An angle of 180° formed by a straight line.
Reflex Angle	An angle which is greater than 180°, but less than 360°.
Vertex	The point at which the two arms of an angle meet.

Perpendicular	A straight line, 90° to another straight line.
Figure	A plane figure is bounded by lines or sides.
Square	A four-sided figure with four 90° corners.
Quadrilateral	A four-sided figure, not necessarily with 90° corners.
Polygon	A regular or irregular figure bounded by more than four sides.
Circle	A continuous line, equal distance at all points from a common centre.
Ellipse	A continuous curved line having two diameters called axes (axis = singular).
True Ellipse	An ellipse that has two focus points (focii) of which the sum of the distances from the focii to any point on the curve is always constant. An inclined cut through a cone or cylinder is a true ellipse.
Approximate Ellipse	An ellipse that is elliptical in appearance, but does not have true focii.

Symbols are often used in geometry to condense wording. The following are common:

Angle	$\angle$	The arc whose endpoints are A and B	$\frown{AB}$	The distance between points A and B; a number	AB
Circle	$\odot$	Is perpendicular to	$\perp$	Therefore	$\therefore$
Parallel	$\|$	Square	$\square$	Equals (two quantities are the same)	$=$
Similar	$\|\|\|$	Triangle	$\triangle$	Is not equal to	$\neq$
Congruent	$\equiv$	Parallelogram	▱	Is greater than	$>$
Plus or Minus	$\pm$	Rectangle	▭	Is less than	$<$
Right Angle	∟				

Unit 1
Straight Lines

1. To bisect any given straight line:

Draw a line AB 3" long, set compass to a distance greater than half of AB (about 2"). With instrument point on A, make an arc above and below line AB. With the same setting, use B as a centre to scribe arcs to cut the first arcs at C and D. With straight edge and pencil, join C to D. This line cuts AB at E and is a true bisection, making AE=EB and a 90° angle.

It is suggested here that the student practice bisection on various lines.

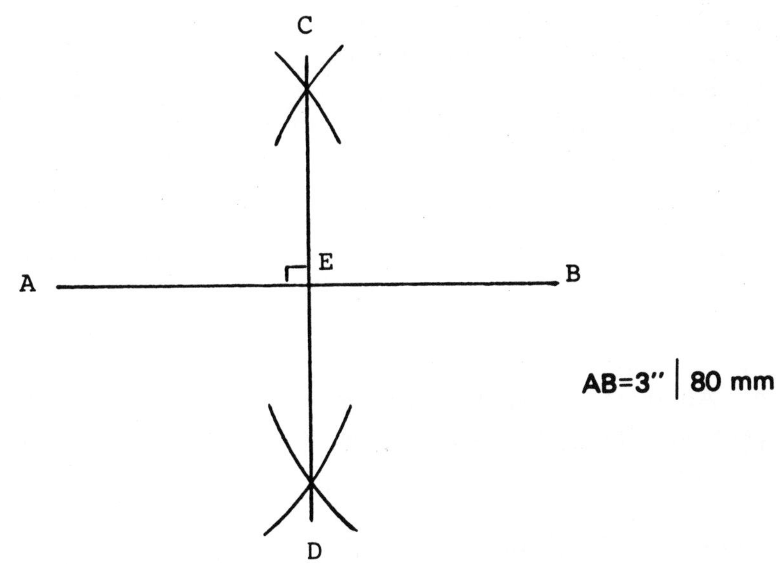

2. To draw a perpendicular to a line, from a given point away from the line:

Draw AB 2½" long. Place P 1¼" away from the line about in the middle. Place compass point on P and swing an arc to cut line AB at C and D. With compass point on C and a radius greater than half CD, make an arc below line AB. Then using D as the next centre, make an arc to intersect the first arc at E. Joining E to P gives the desired perpendicular, which is 90° to AB.

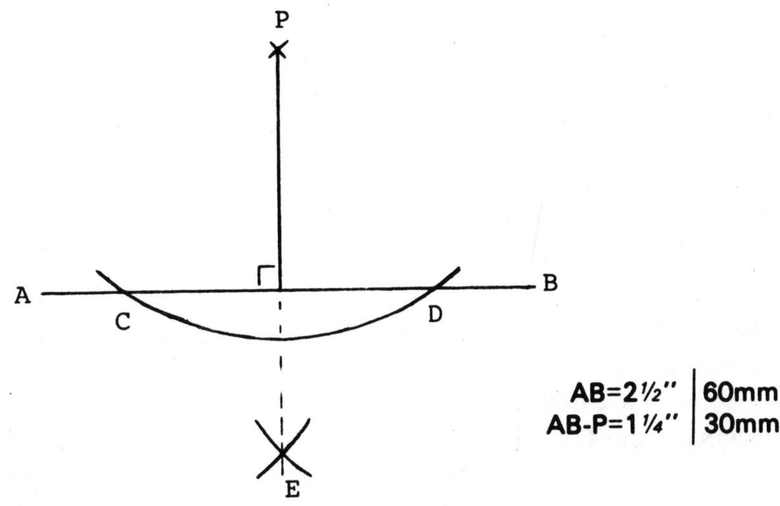

3. To draw a perpendicular at the end of a line:

Draw line AB 3¼" long, use B as a centre point and a radius of about 1/3 of AB. Scribe an arc; almost a semicircle. With the same radius, step off arc D from C and arc E from D. Using D and E as centres and the same radius, scribe arcs to make F. Join F to B. FB is perpendicular to AB.

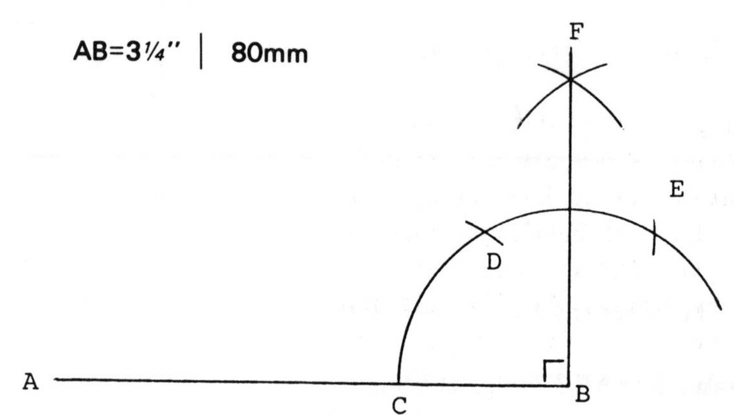

4. Alternate method of drawing a perpendicular from a given point on the line:

Draw line AB 3½" long, place P ¾" in from B. Select C anywhere above line AB. Use CP as a radius and C as centre to scribe an arc that cuts AB at D. Draw a line through CD to cut the arc at E. Connect E to P thus constructing a perpendicular to AB.

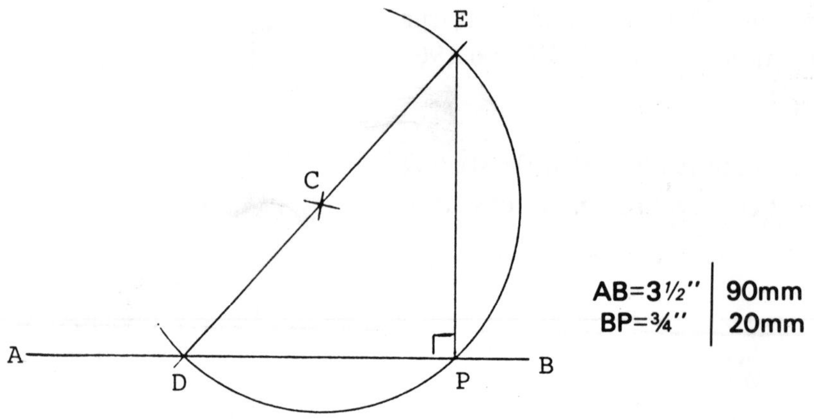

5. To find a point on a line equal distance from two established points: CD

Bisect points CD as you would bisect a straight line. Let the bisection line cut AB at E. E is equal distance from C as from D.

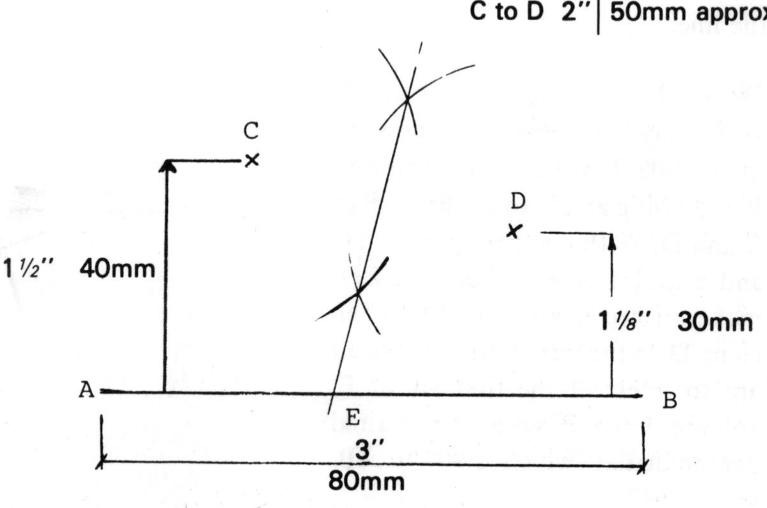

6. To layout a parallel line:

Using the given distance, say 1¼" for a radius, scribe two arcs from AB close to each end. Use a straight edge and draw the line CD just touching both arcs. CD is a tangent to the arcs and is parallel to AB. This tangential method is good for many situations, but not as accurate as Number 7.

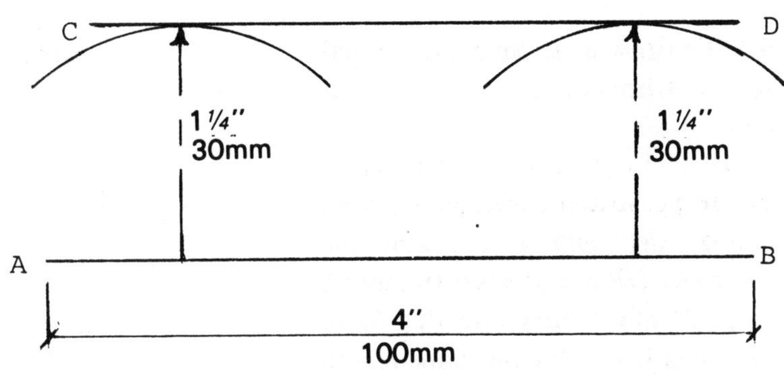

7. To make a parallel line through a given point: C

Place C <u>about</u> 1" away from AB and in from the end. From C make an arc at D using any reasonable radius. Use the same radius from D to swing arc FC. With compass make DE the same length as FC and join points C and E.

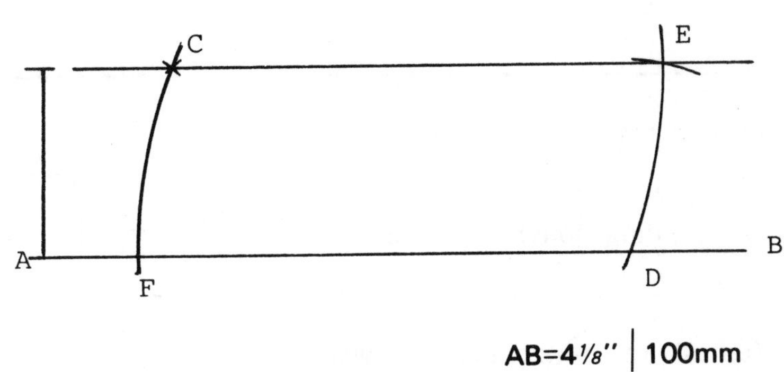

8. To divide a given line into any number of equal spaces:

Make AB the given line and draw AC at any reasonable acute angle from A. On AC, mark off the required number of equal spaces by measure or compass. Join 5 to B, then lay a set square along line 5B and a straight edge along the base of the set square. Now slide the square and join points 4, 3, 2, 1 to line AB thus keeping them parallel to 5B. AB is now divided into five exactly equal spaces.

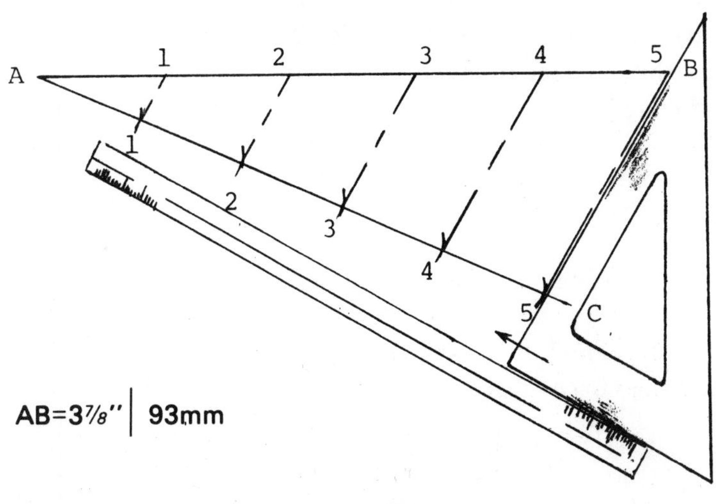

9. To divide a line into equal spaces with compass and straight edge only:

Make AC an acute angle to line AB. Scribe arc DO any reasonable distance from A. Use the same radius from B to scribe EO. With compass make EO the same length as DO. Now draw line BF through E. Divide AC into the given number of spaces, say five again, and BF the same. Next join points A to 5, 1 to 4, 2 to 3, 3 to 2, 4 to 1, and 5 to B.

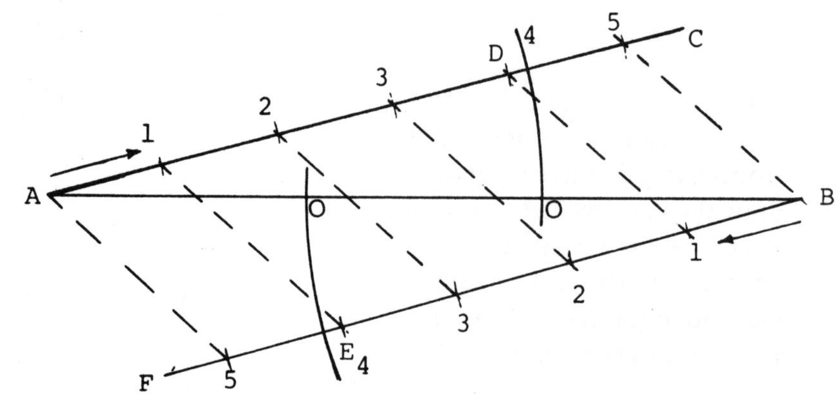

AB=5⅛" | 130mm

10. Finding louvre spacing.

Vertical or horizontal, for wall dividers or ventilation, louvres should overlap to block direct vision. Within the space to be divided, layout the top and bottom louvre at 45°. Select a suitable round figure (X)", less than (Y)", therefore, Z" over X" = number of spaces. Now geometrically divide Z by this number to layout the exact spacing W.

Scale 1:5

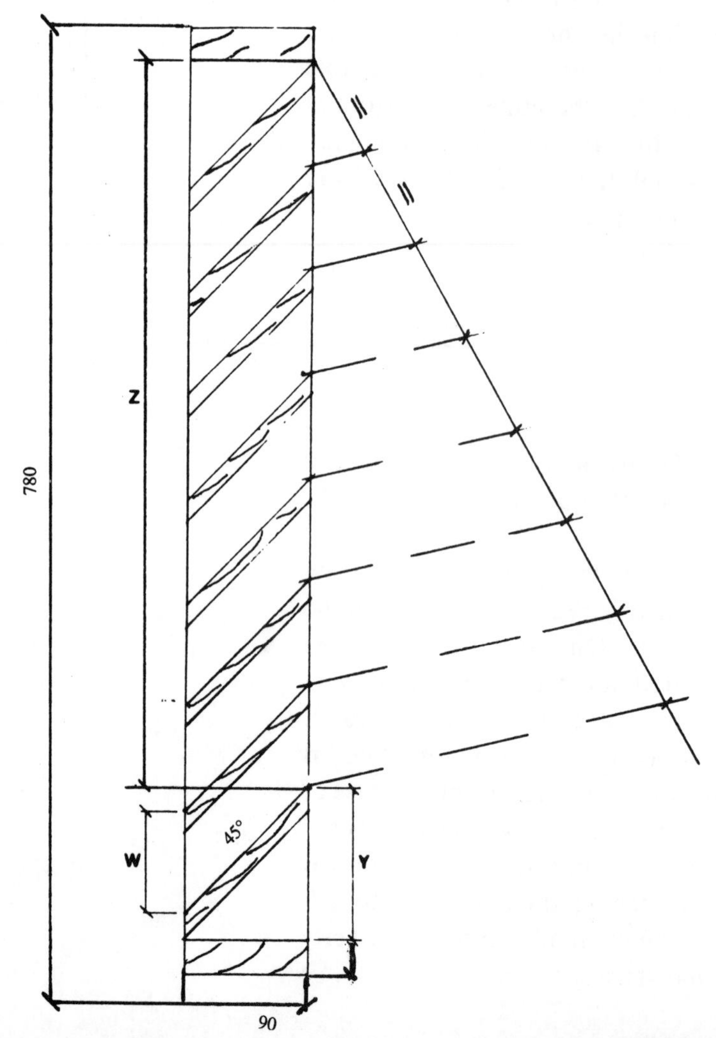

Discovery Exercises

1. Use a set square to draw a line 30° from the horizontal. Make the line 2¾" (70) long and bisect it.

2. <u>Without</u> a set square, layout on a base line a square with 2¼" (55) sides.

3. Use an acceptable method to layout a perpendicular to line AB from point C.

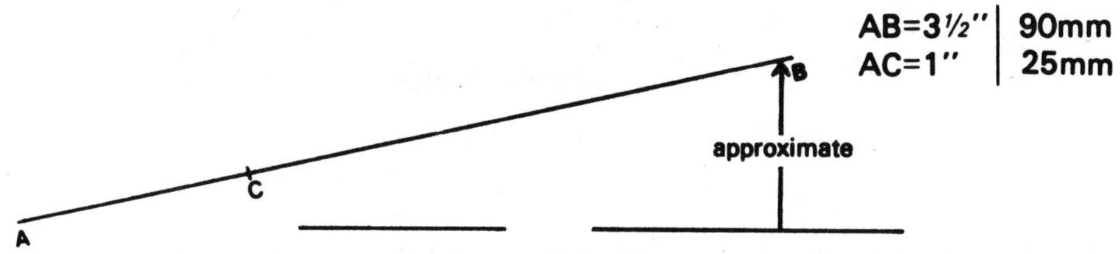

4. From points C and D, drop perpendiculars to line AB. Then from C and D, layout two lines of equal length meeting on line AB. Do not extend line AB at B.

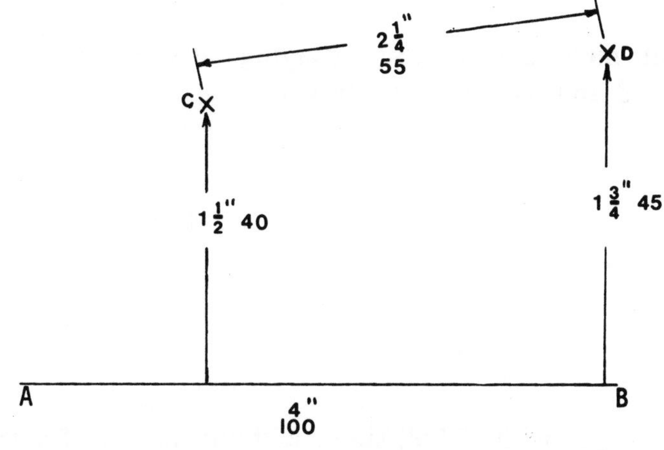

Discovery Exercises

5. 5' (1500) away from, parallel to the existing wall and starting at wall AB, layout a 9' (2700) long line and geometrically divide it into 7 equal parts.

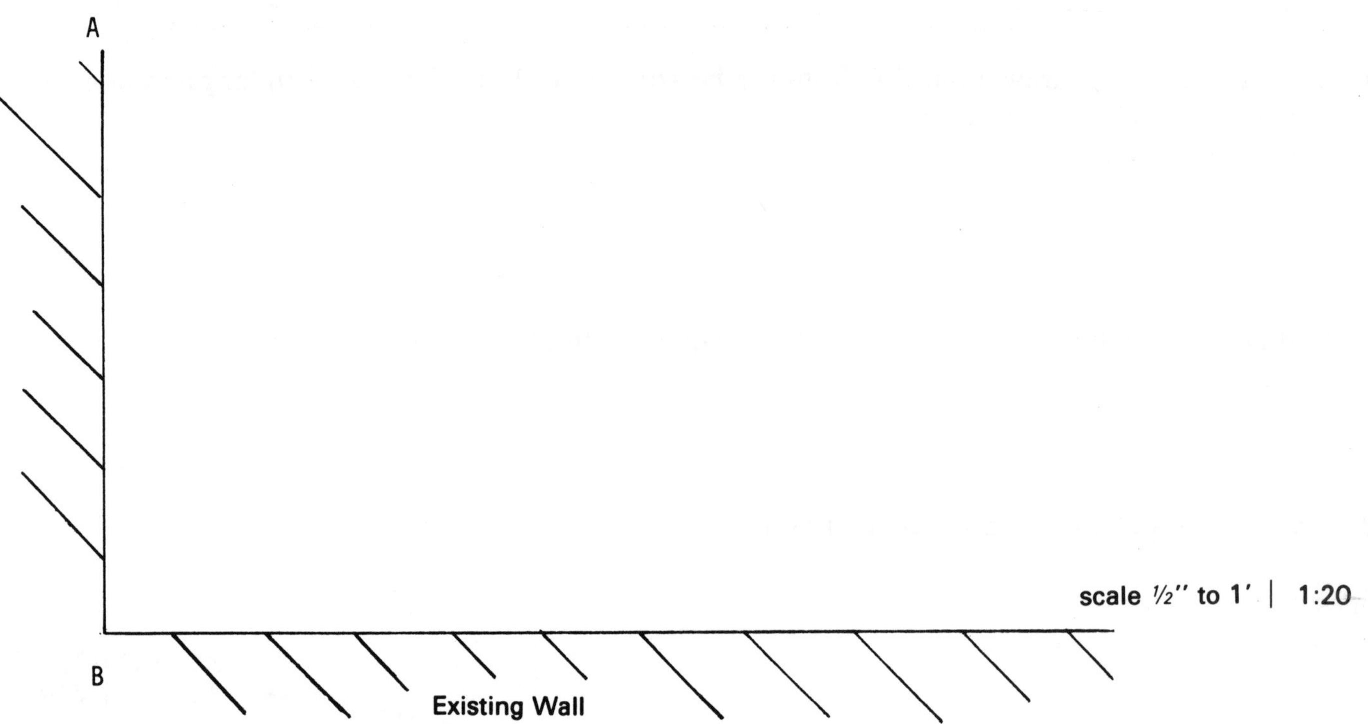

6. Find an alternate method to exercises 1 to 4 of laying out a perpendicular at P when AB is 3¾" (95) and P is ½" (14) from B.

7. Find an alternate method to exercise 2 of laying out a perpendicular to line AB through P when AB=4" (98) and P is located approximately as shown.

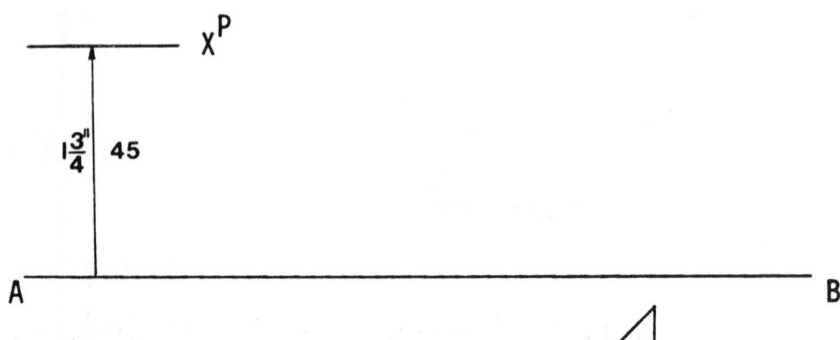

8. Layout a 45° ∟ △ with 2" (48) sides, without the aid of a set square.

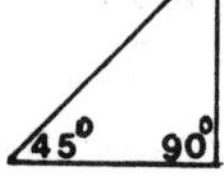

Unit 2
Angles and Triangles

A triangle is a figure contained by three straight lines.

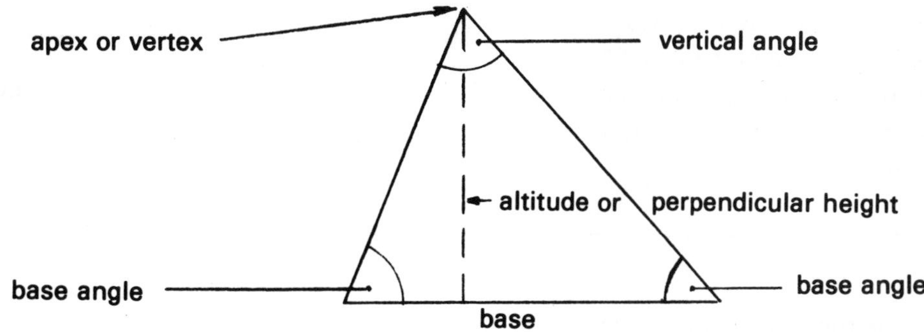

Triangles are named either for the comparative lengths of their sides, or for the sizes of their angles.

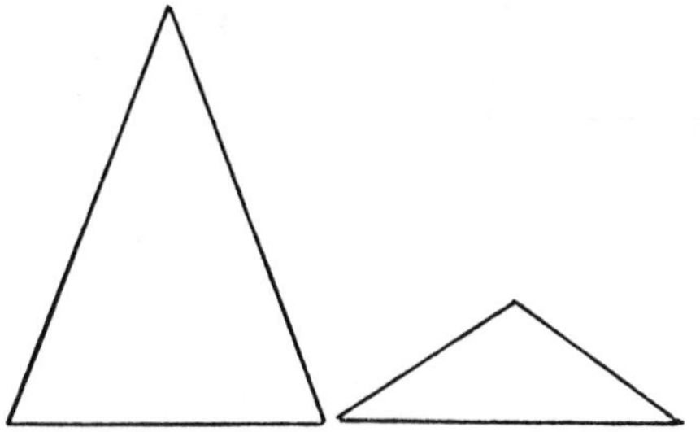

An isosceles triangle has two of its sides and two of its angles equal.

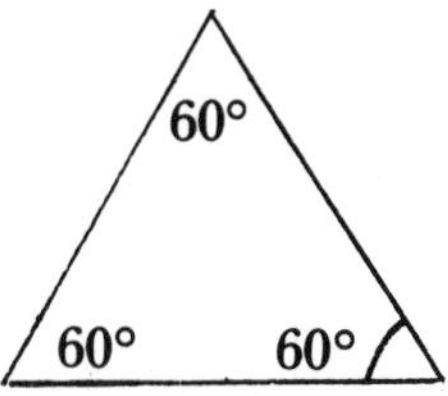

An equilateral triangle has its three sides and angles equal.

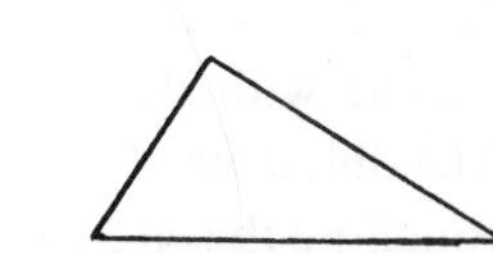

A scalene triangle has three unequal sides and angles.

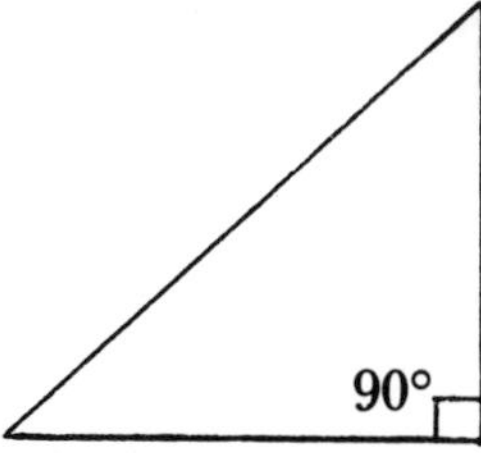

A right-angled triangle has one of its angles 90°. The side opposite is called the hypotenuse.

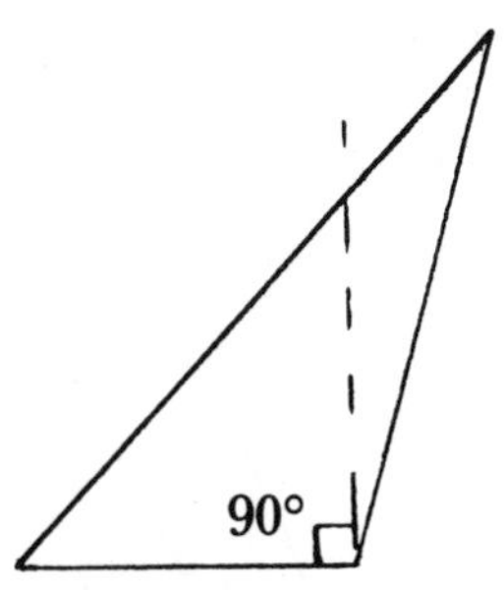

An obtuse-angled triangle has one of its angles obtuse; greater than 90°, less than 180°.

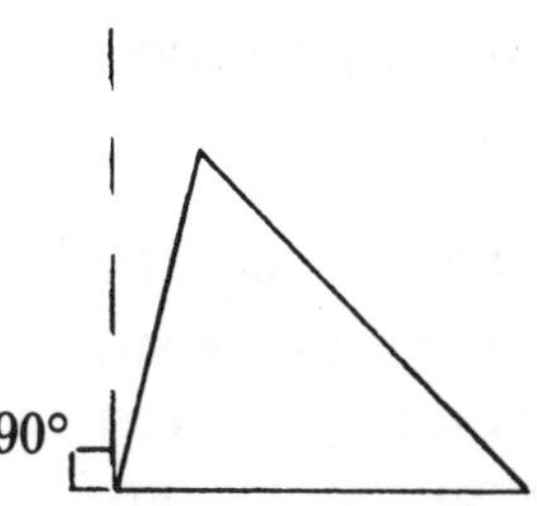

An acute-angled triangle has all its angles acute; less than 90°.

Angles and Triangles

The internal angles of all triangles always total 180°.

1. To construct a right angle triangle using the 3 4 5 method:

This method is commonly used in building layout. On a base of 4", use B as a centre to strike arc C 3" away. Use A as a centre and with a 5" radius, cut arc C. Note the figures 3 4 5 may be increased proportionally to 6 8 10, etc.

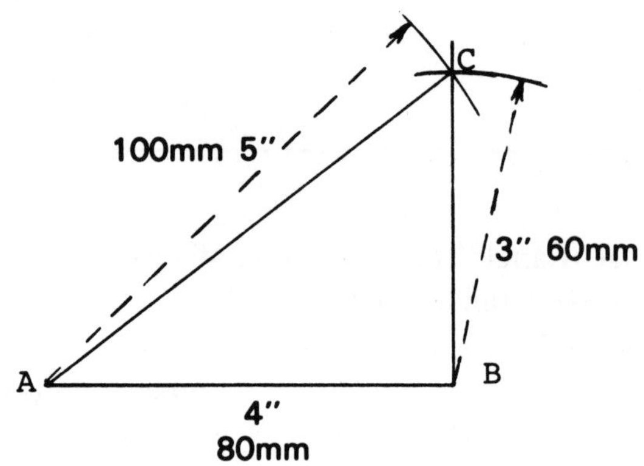

2. To construct a right angle triangle given its hypotenuse and one side:

Make AB 3" long, bisect AB at O and scribe a semicircle using OA as radius. With centre B and the radius of given side 1⅛", scribe an arc to cut the circumference at C. Join A to C and B to C. It is wise to remember that a triangle with its hypotenuse equal to the diameter of a circle and the heel touching the circumference, is always a right angle triangle. (See Straight Lines, problem 4.)

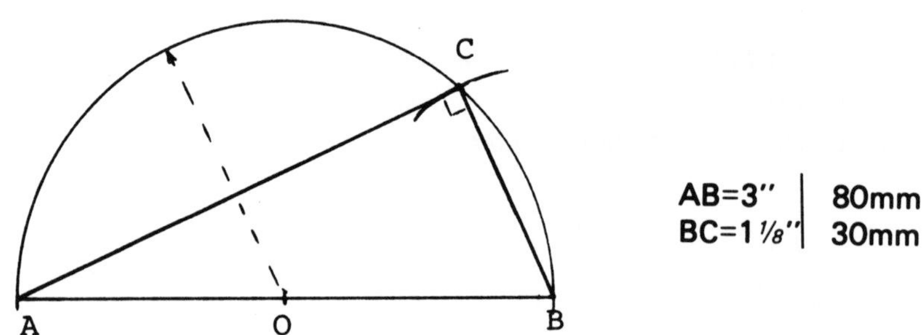

3. To construct an equilateral triangle:

Make AB 2½" long. Use A as centre and AB as radius to strike an arc at C, then use B as centre and BA as radius to cut the arc at C. An equilateral triangle can also be drawn with a 60° set square; thus proving all its angles add up to 180°, which makes it also equiangular.

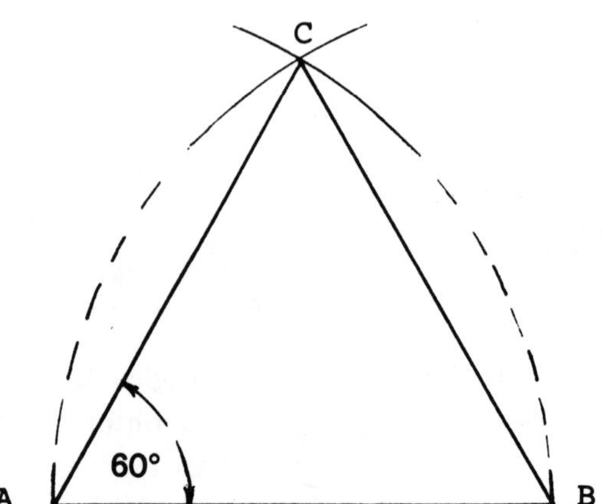

AB=2½" | 60mm

4. To bisect any angle:

From the apex A, use any reasonable radius to scribe arc CD. Using C and D as centres and a reasonable radius, strike intersecting arcs at E. Join A to E. It should be remembered that an angle is always represented by the centre of three letters; CAB ∴A=the apex.

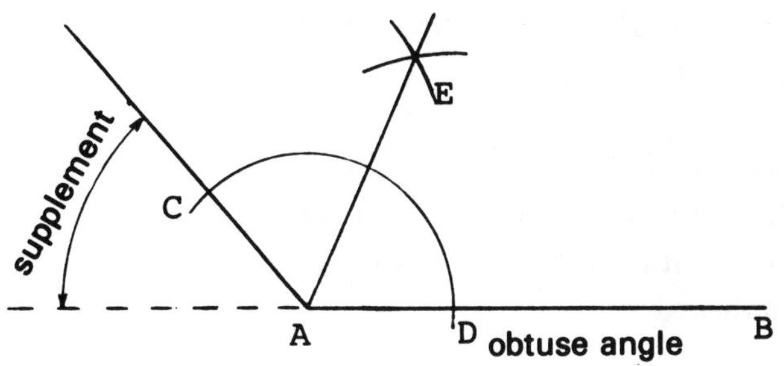

NOTE: To transfer an angle to another line, redraw arc CD using AD as the radius, then with a compass, measure arc CD. A line is then drawn through AC.

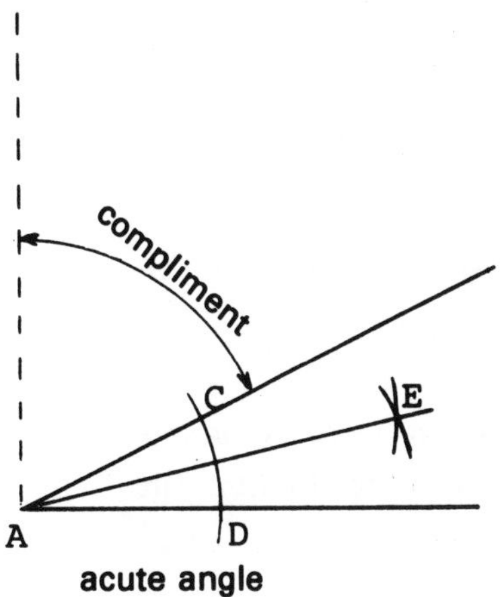

5. To trisect is to divide into three equal parts:

Make a 90° angle. From A, strike arc BC. Use this same radius from B to cut at D, then use the same radius from C to cut at E. Connect A to D and A to E. This is to be remembered for making 30° and 60° angles.

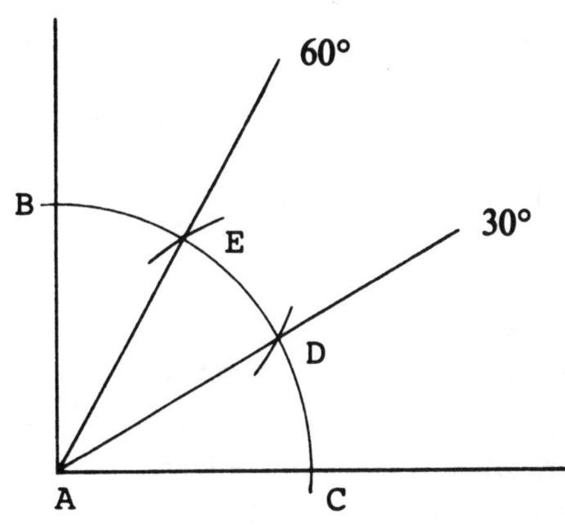

6. To make many other angles such as in a protractor:

Bisect a 90° angle and trisect it. Then continue to bisect each new angle until sufficient angles are established.

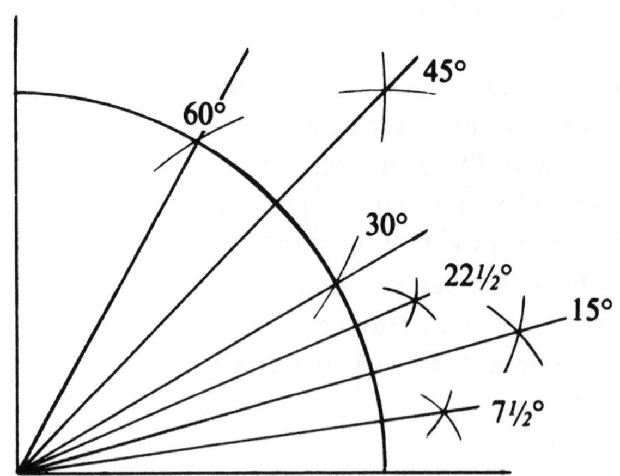

7. To construct an isosceles triangle given the vertical height and the base:

Erect a perpendicular at the bisection of a 2½" base line. Mark off the height 1⅛" at C and connect C to A and C to B.

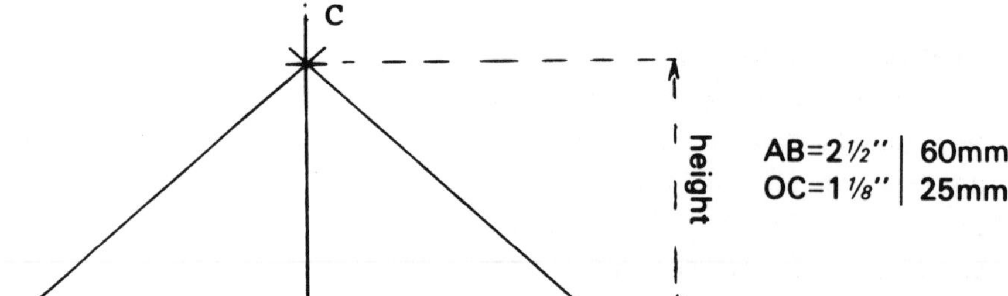

8. To construct a triangle given the length of each side:

Draw base AB 2½". With centre A and radius AC, scribe an arc. With centre B and radius BC, scribe an arc to cut the first arc at C. Connect A to C and B to C.

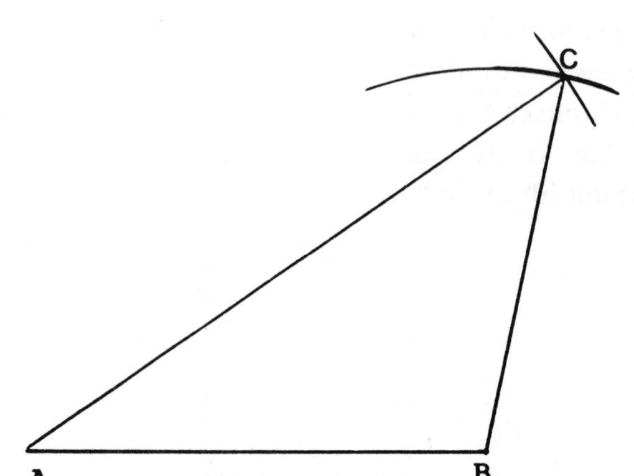

9. To construct a triangle given one angle, one side and the perimeter:

Draw a line 1¼" as base AB and lay off an angle of 45° at ABC. Mark off B to C = to the perimeter of 5" minus base AB ∴ BC =3¾". Connect A to C. Bisect AC and let the bisection line cut BC at D. Connect D to A to form the desired triangle.

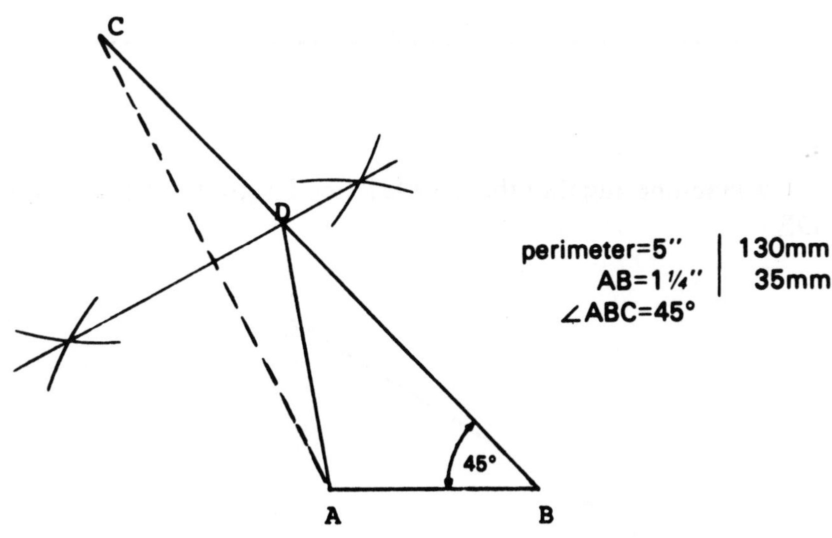

10. Finding unseen shapes.

To build this mould, surfaces X and Y must be discovered. Given some information such as 1-2, 3-4, and angles, find X by using D and E about 90°. To find Y, use G and F about 90°. Other required shapes are readily seen.

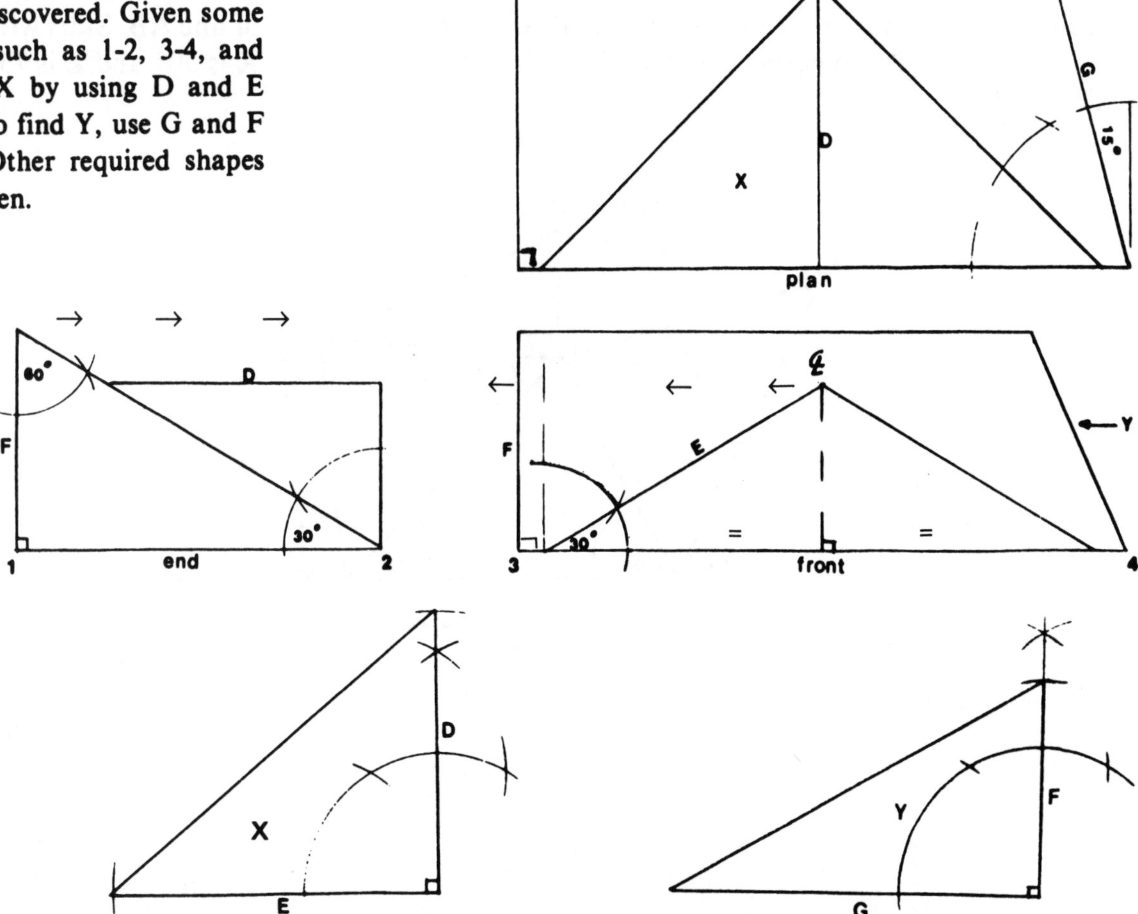

Discovery Exercises

1. Remembering that the middle letter designates the angle referred to, reproduce ∠BAC as DEF on line DE.

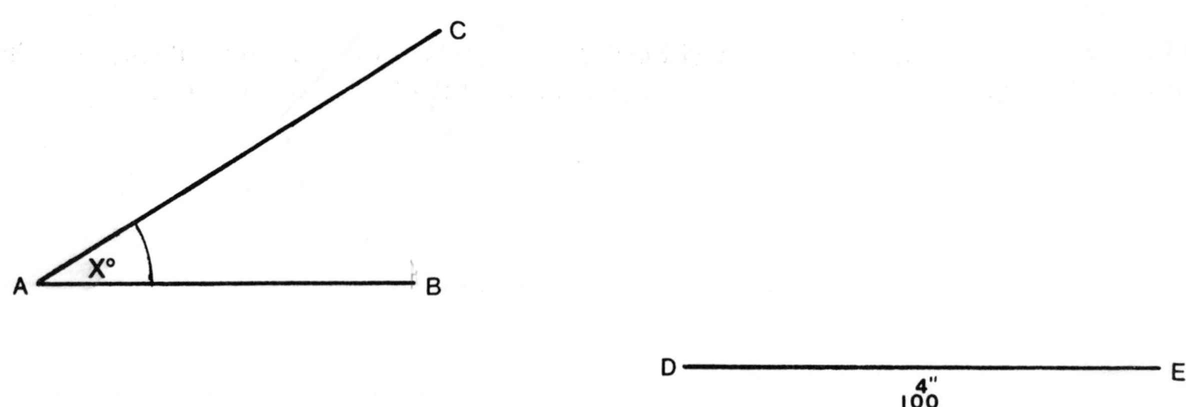

2. Accurately reproduce the figure shown, and state the exact length of line AB. Start with base line 2½" and <u>show all compass construction of angles</u>. Also, if internal angle X is 50° and Z is 75°, what does Y equal?

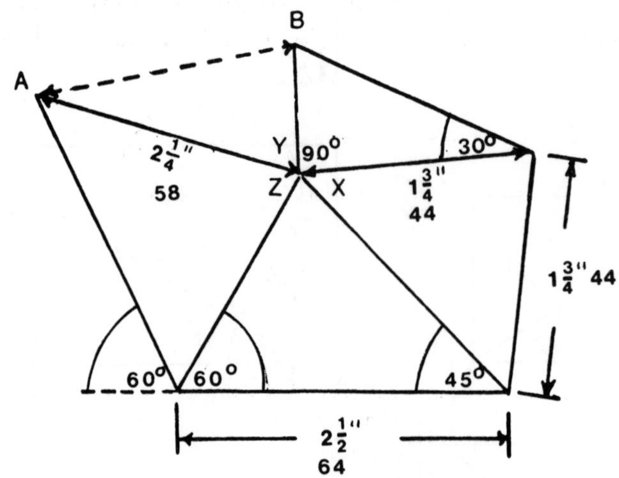

3. Layout the figure shown to full size.

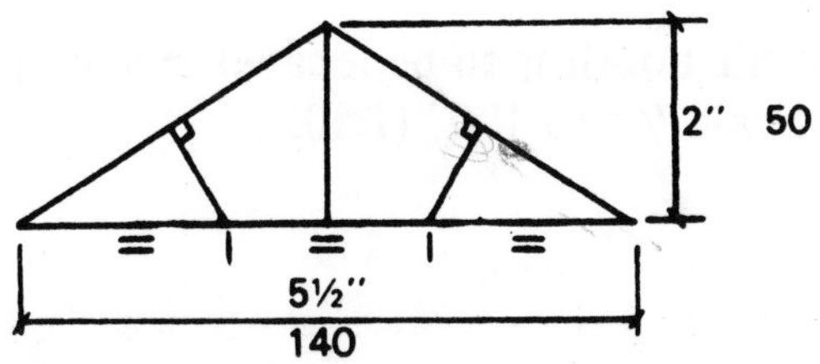

4. ABC is an obtuse angle, and X is a point within it. Determine X's position when AB=2⅜" (60), BC=3¼" (83), ABC=150°, ABX=75°, and BCX=22½°. State the lengths of AX, BX, and CX.

5. A radial arm saw has to be set to precut mitres on moulding to fit around an object which has obtuse and acute angles. Show on a drawing the correct angles to which the machine should be set.

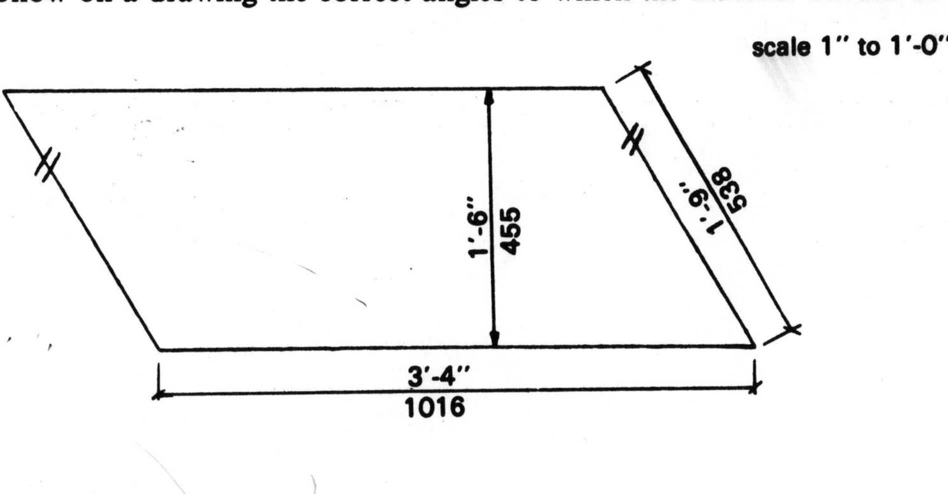

6. A souvenir booth is to be located in a shopping mall. Layout the floor plan as shown by geometric methods. Scale ¼" to 1'-0" (1:50).

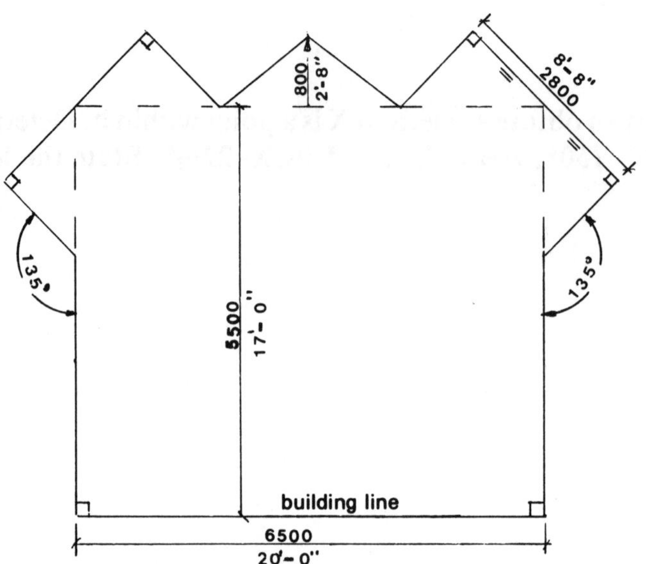

building line=exterior wall surface

Unit 3
Quadrilaterals

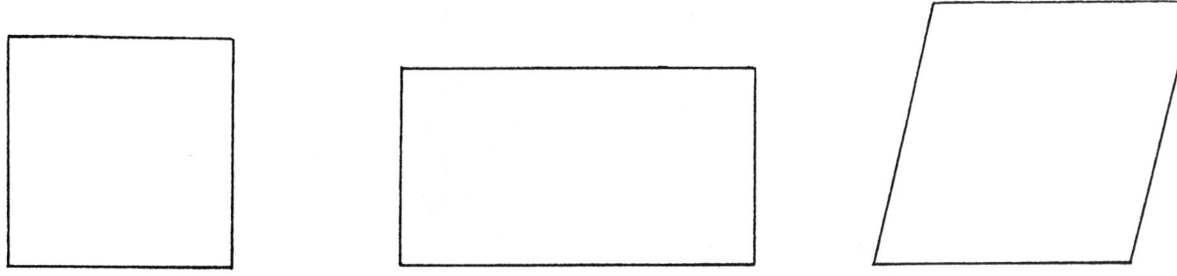

A square has equal sides and all right angles.

A rectangle or oblong has its opposites sides equal and all right angles.

A rhombus has all its sides equal, but its angles are not right angles.

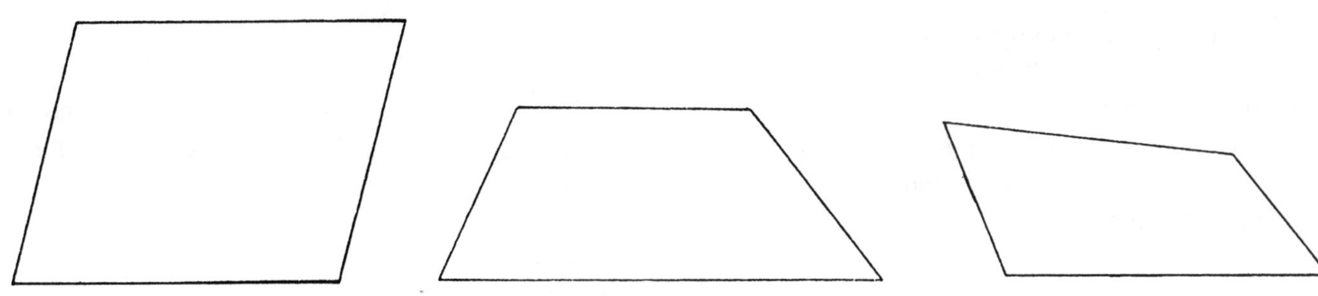

A rhomboid has its opposite sides equal, but its angles are not right angles.

A trapezoid has only two of its sides parallel.

A trapezium has none of its sides parallel, but may have two of its sides equal. When two of its sides are equal, the figure is sometimes called an isosceles trapezium.

Quadrilaterals

1. To construct a square given its diagonal: 3"

Draw the diagonal AB 3" long and bisect it. Using O as centre and OA as radius, scribe a circle. Where the circle cuts the bisection line, connect CD to A and B.

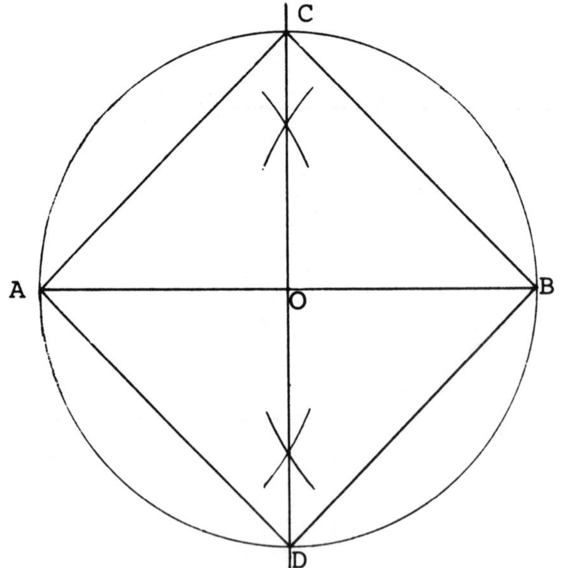

AB=3" | 80mm

2. To construct a rectangle given the diagonal and one side:

Draw a circle with the diagonal as a diameter. From A and B, mark off the length of the given side on the circumference and connect points A, D, B, C. (See Angles and Triangles, problem 2.)

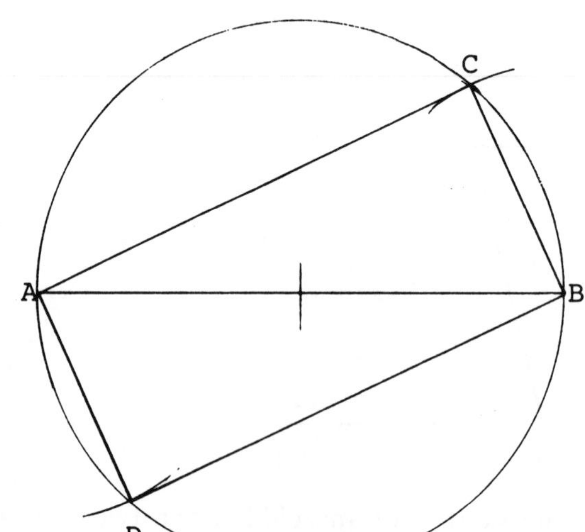

AB=3¼" | 80mm
AD=1¼" | 30mm

3. To construct a quadrilateral with one 90° corner, given four sides:

Draw AB to one given length and erect a perpendicular at B. (See Straight Lines, problem 3.) Using a compass, mark off one other given length at C. From C, mark off one more length with an arc at D. From A, mark off the last length, to cut the arc at D. Connect AD and DC.

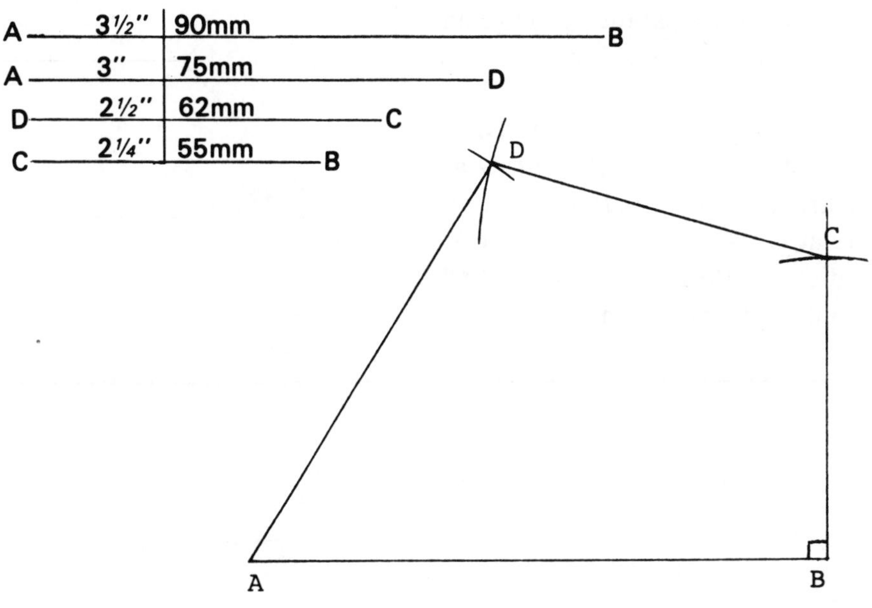

4. To reduce or enlarge a rectangle:

Draw a rectangle ABCD 3¼" x 1½" and make one diagonal. Given one side of the new rectangle 1", at AE and BF, join E to F and drop a perpendicular to AB from 0. (See Straight Lines, problem 2.) The long side of the new rectangle can now be measured. To enlarge, extend lines AD, AB, AC then proceed as above.

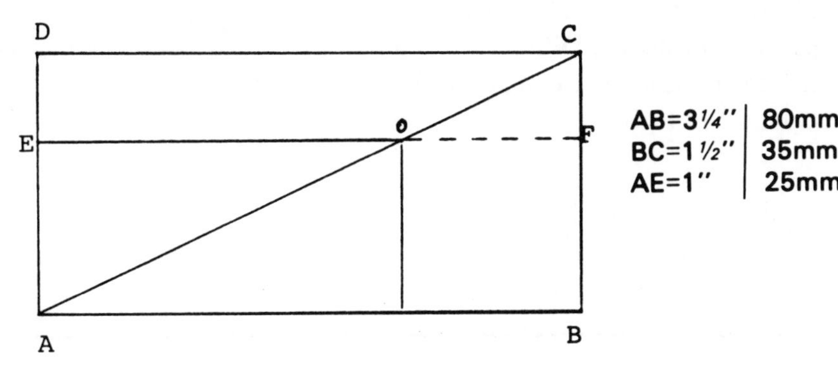

5. To make a parallelogram from an equilateral triangle:

Layout the triangle with 2¼" sides. With B as centre, scribe arc ACD, then lay off CD equal to AB. Connect C to D and D to B. This figure is called a rhombus.

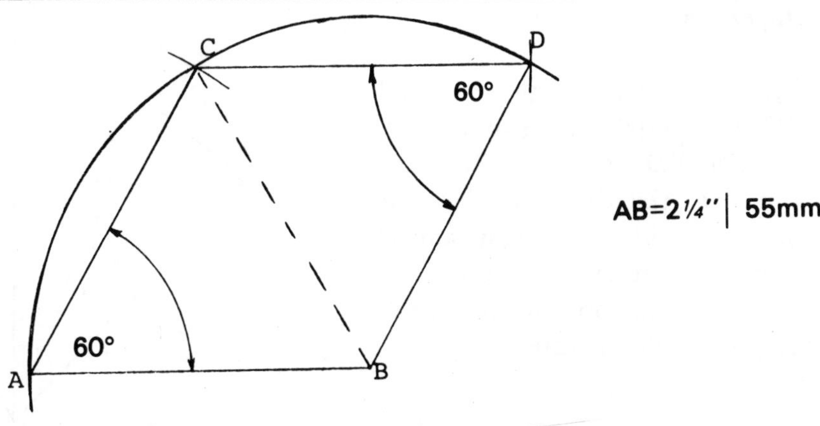

6. To make a parallelogram given one angle and two sides:

Draw AB 3" and lay off angle DAB 45°. Measure AD 1¼". With D as centre and AB as radius, scribe an arc at C. With B as centre and 1¼" radius, cut the arc at C. Connect all points to form a rhomboid.

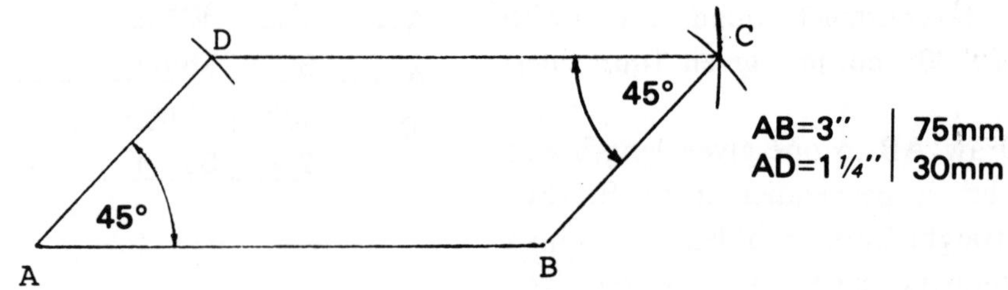

7. To layout a trapezoid given the length of two parallel sides and one other side adjacent to a given internal angle:

Draw AB 3¼" long. With A as the vertex, construct a 60° angle and measure off AC 1¾". Draw a parallel through C. With 1⅞" as a radius and C as centre, mark off D. Connect D to B.

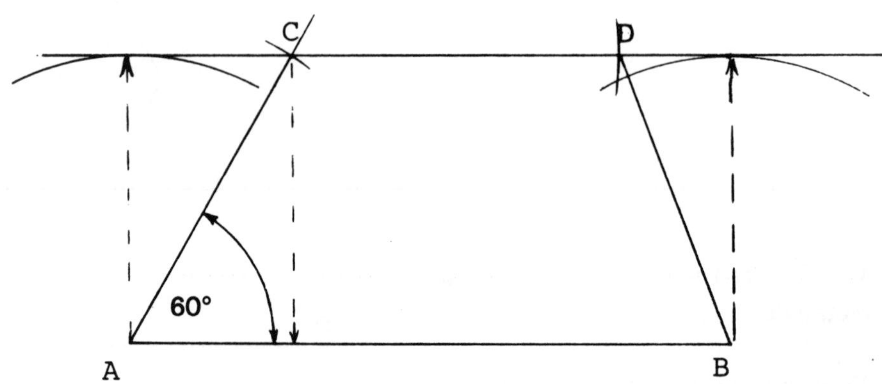

AB=3¼"	80mm
CD=1⅞"	50mm
AC=1¾"	45mm

8. Using triangulation to layout a trapezium:

Set off on a straight line AB 1", and BC 1¼". Perpendicular to this line, set out BD ¾", and BE 1¼". Connect ADCEA. The figure is made up of four right angled triangles. A reversal of this method may be used in breaking down odd shapes to calculate area.

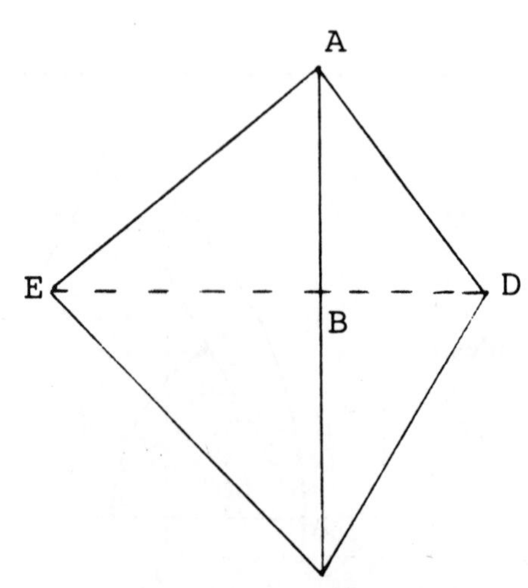

AC=2¼"	55mm
ED=2"	50mm
BA=1"	25mm
BC=1¼"	30mm
BD=¾"	20mm
BE=1¼"	30mm

9. This problem is similar to number 8. Certain dimensions must be known to layout irregular quadrilaterals. This time A to D is the first measured line.

Use a scale of ¼" to 10'- 0". AB 100', BC 100', CD 80', CF 70', BE 90'. This is how a surveyor might record the data of a survey, the area could now be easily calculated.

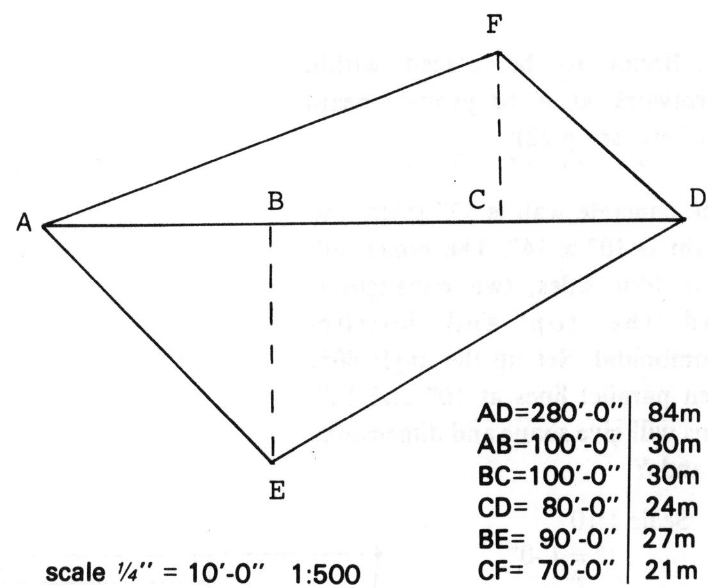

10. To relate quadrilaterals to work in a later unit on intersecting planes, attempt to solve this problem:

A chimney 18" (460) square, passes through a roof sloping at 30°. Discover the shape of the hole which must be cut in the roof sheathing.

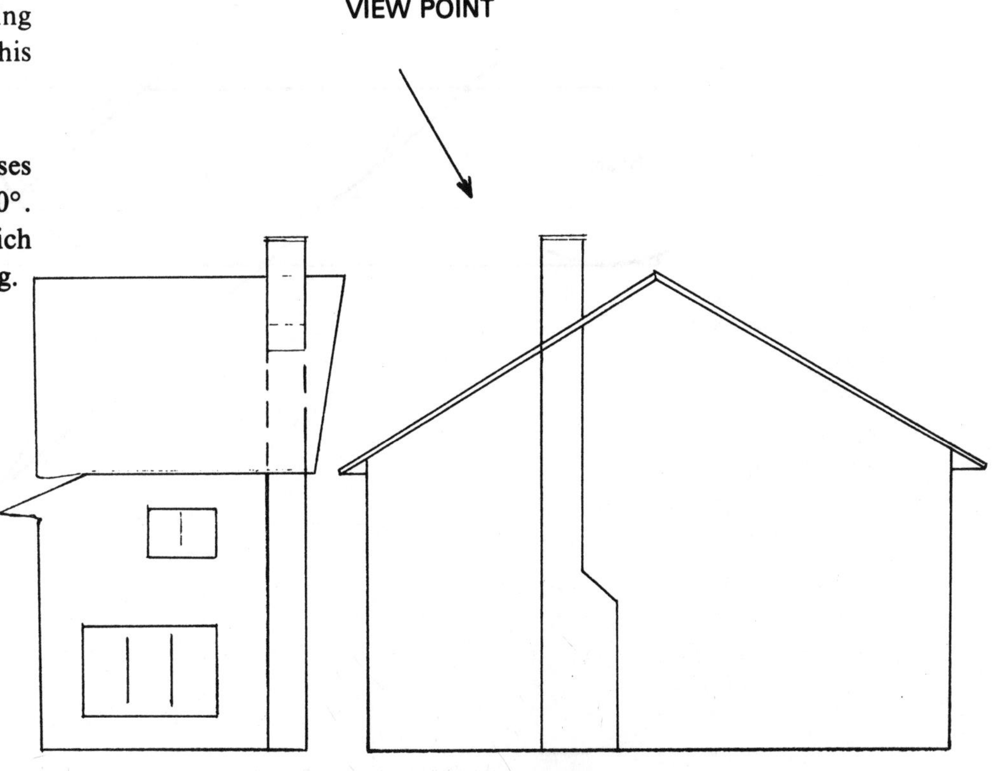

11. Boxes to be placed within formwork at A to provide beam pockets (see p.22):

The concrete wall is 12" thick, the beam is 10" x 14". The boxes will have four sides, two rectangular, and the top and bottom rhomboidal. Set up the angle 40°, then parallel lines at 10" and 12". This will give shape and dimensions X and Y.

Scale 1:10
1" = 1'-0"

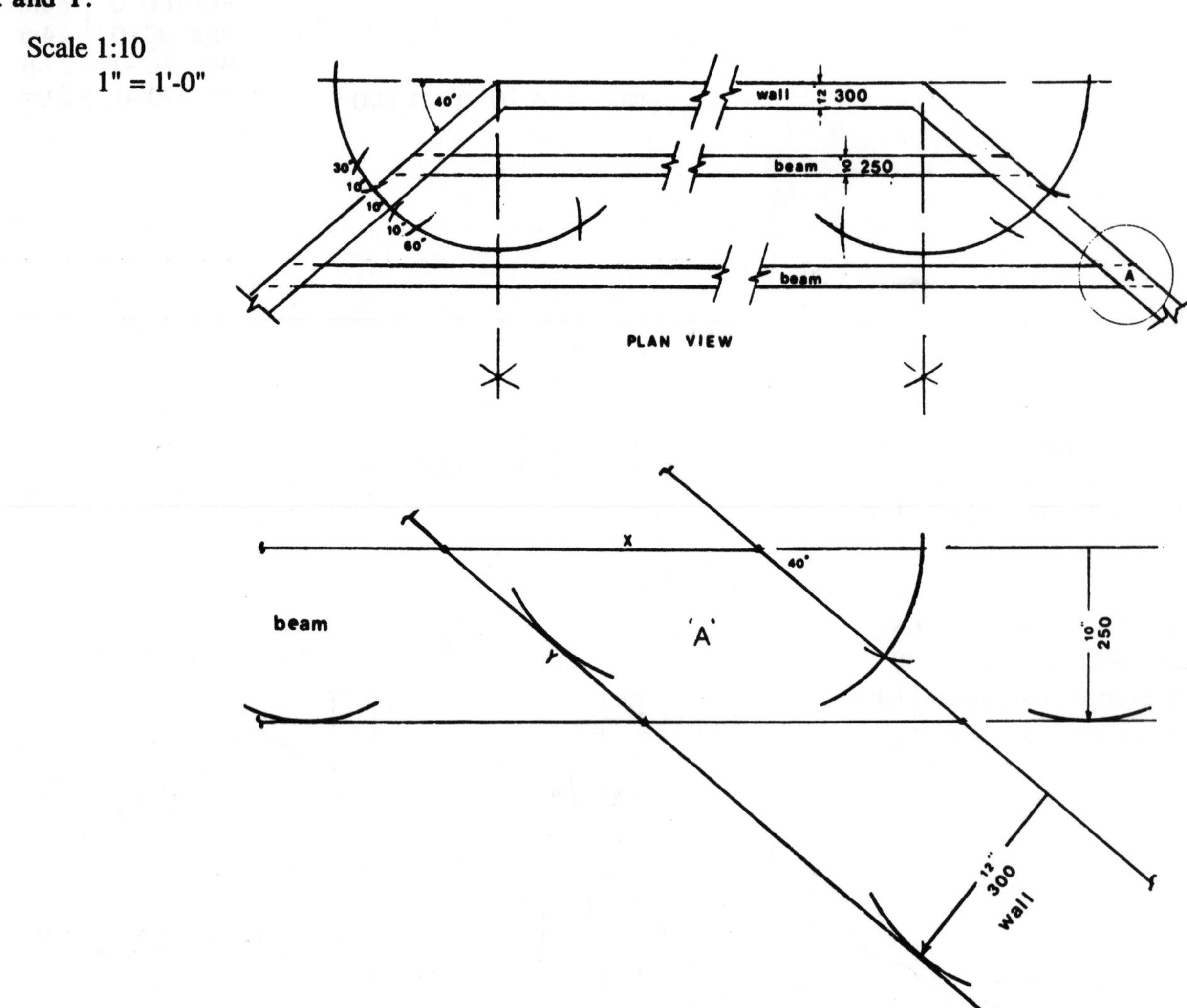

Discovery Exercises

1. A building is to be laid out as per drawing. Redraw the layout to a scale of 1/4" to 1' (1:50) and show geometrically how the figure might be broken down into the least number of right angle triangles for area calculation. Do not calculate the area. Show construction lines.

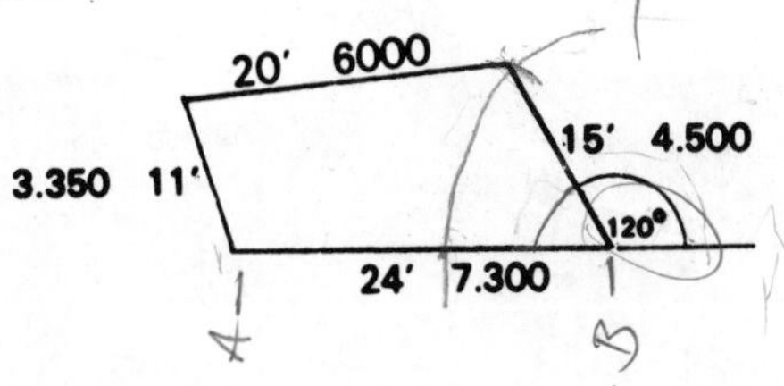

2. Place the given series of right triangles into a quadrilateral, with the hypotenuse of each making up the perimeter. State the length of X. The two 30° must form an apex of 60°.

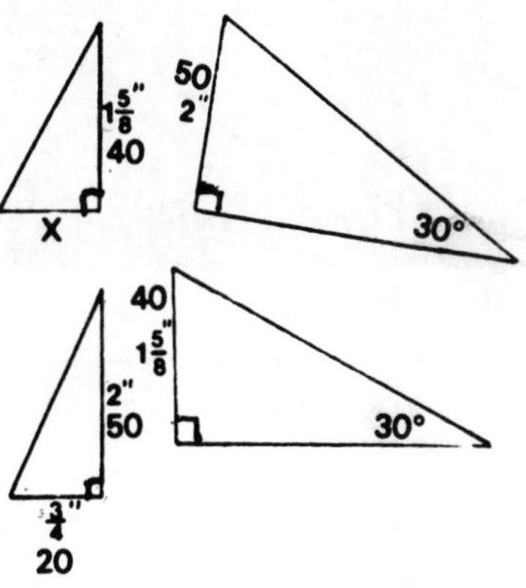

3. Draw a quadrilateral on line AB with angle ABC 60°, BC 1 1/4" (32), CD 2 1/2" (64), and AD 1 3/4" (45). Also show an approximate method of finding the centre of this figure.

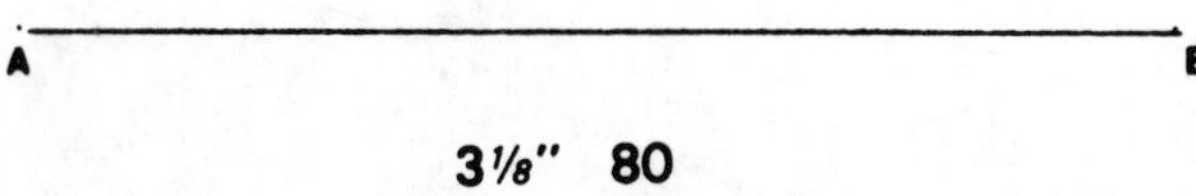

4. Reproduce the given shape to a scale of 1" to 1' (1:10) and layout the position of a hole to be cut in this counter top 6" (150) from, and parallel to, the four sides.

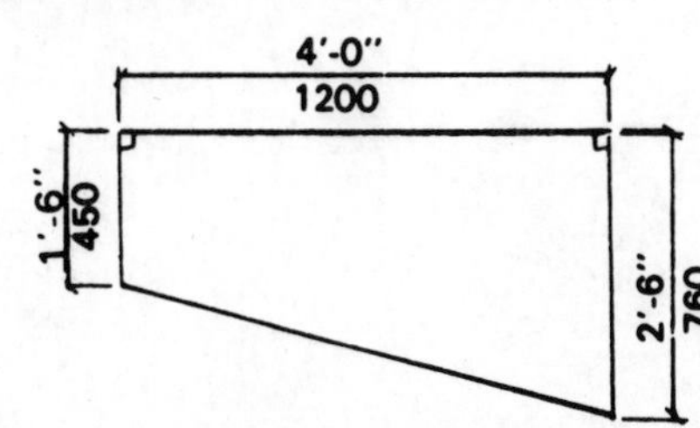

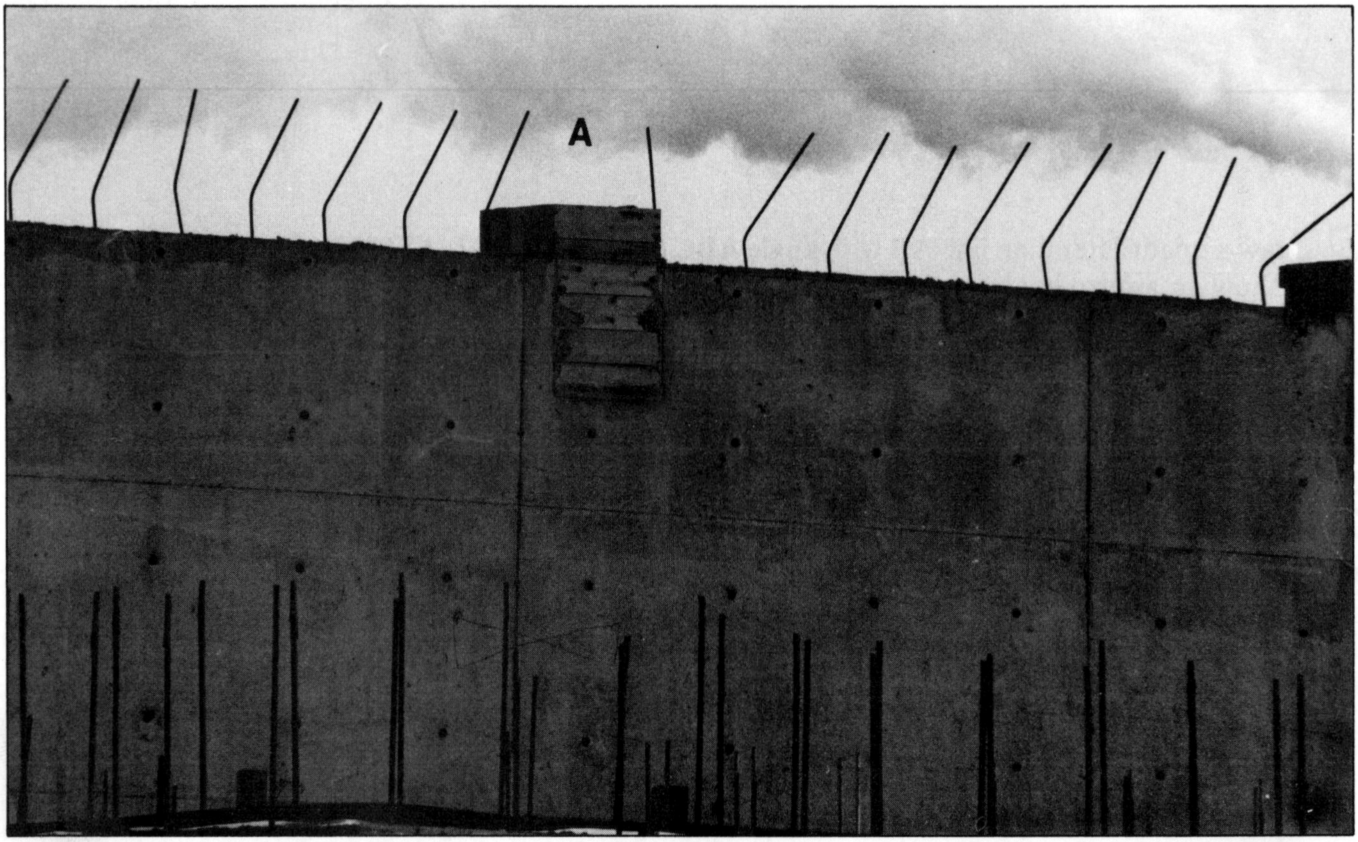

A

24 Circles

1. **The circle and its properties:**

Circumference	The continuous curved line.
Diameter	Any straight line passing through the centre and each end terminating at the circumference.
Radius	Any straight line from centre to circumference (half a diameter).
Chord	A straight line shorter than the diameter touching the circumference at both ends.
Tangent	A straight line touching the circle, but not cutting it.
Normal	A straight line perpendicular to a tangent at the tangent's point of contact with circle.
Arc	Any part of the circumference.
Segment	That portion between a chord and the circumference.
Sector	A pie shape between two radii and an arc.
Quadrant	A quarter of the circle enclosed with two radii at right angles to each other.

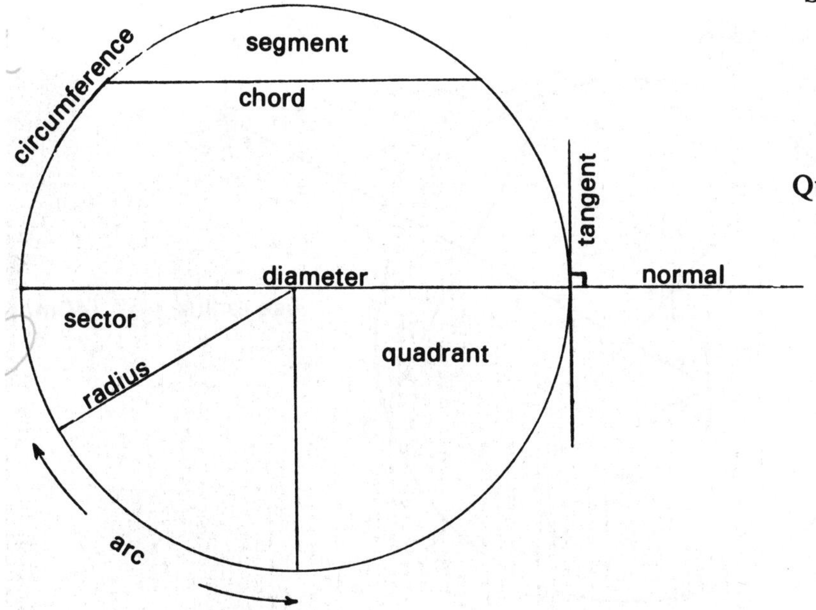

2. Eccentric circles

Circles enclosed within one, each with a different radii and a different centre.

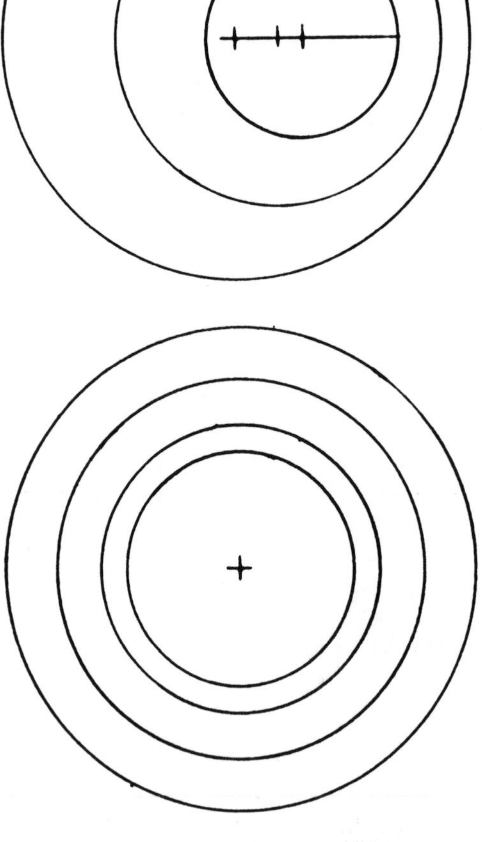

Concentric circles

Circles using the same centre, but different radii.

3. To find the centre of a given circle:

Draw two chords AB and CD, then bisect them. The intersection of the bisectors is the centre O.

NOTE: The closer to 90° the bisectors cross, the easier it is to pinpoint the centre.

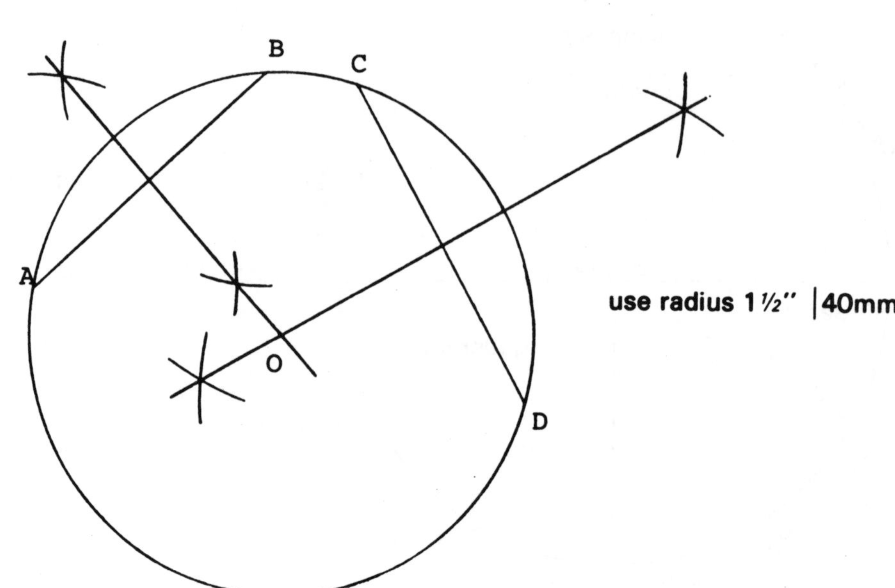

4. To find the centre of a given arc:

Proceed in the same manner as for problem 3.

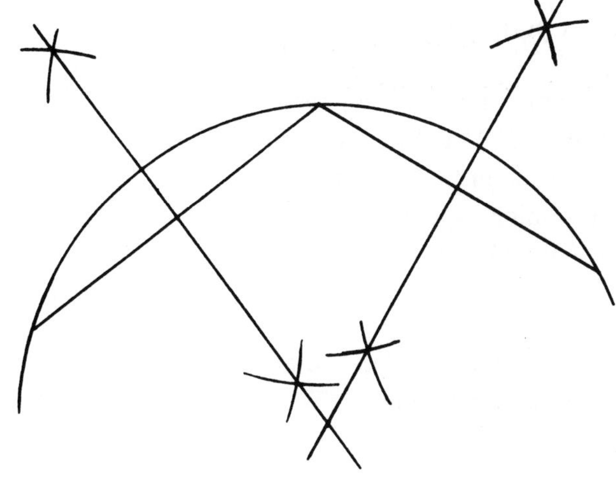

5. To draw a tangent to a circle at a given point:

Produce a radius line through the point P. Construct a perpendicular to OP at P. QR is the required tangent.

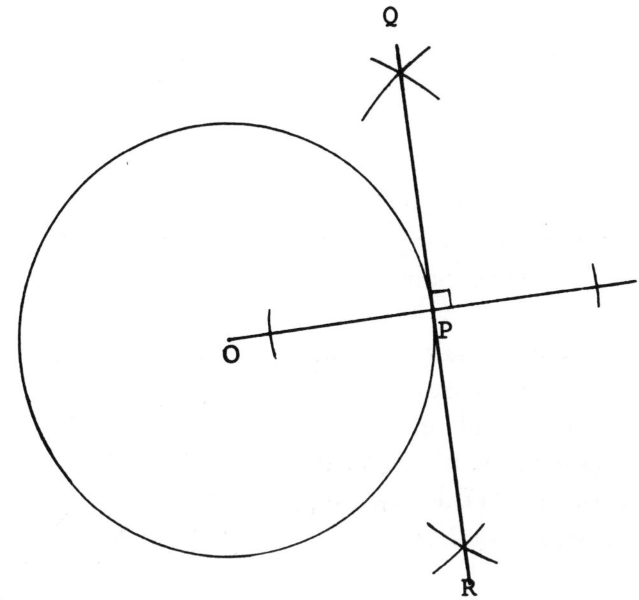

radius=1 ¼" | 30mm

6. To draw a tangent to a circle from a given point away from it:

Produce a radius line to the given point P. Bisect OP. Use the bisection point Q as centre and QP as radius, to scribe a semicircle that cuts the circle at R. Join RP for the required tangent.

NOTE: The angle within the semicircle is a right angle. (See Straight Lines, problem 4.)

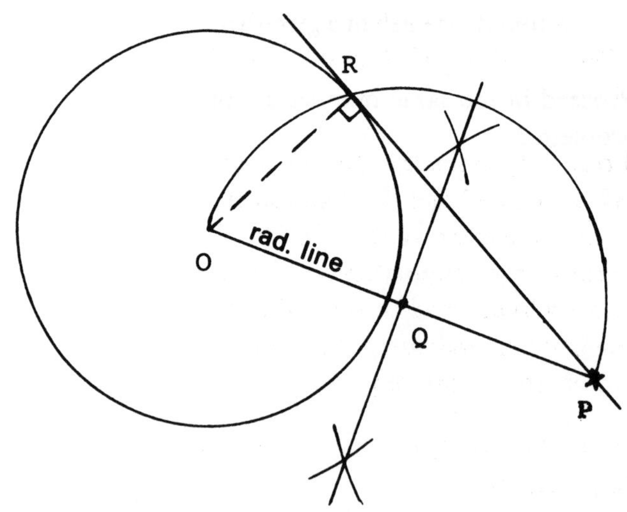

radius=1 ¼"	30mm
OP=approx. 2¾"	70mm

7. To draw a circle of given radius to touch an existing circle and to pass through a given point, P:

Set compass to the new radius and with P as a centre, strike an arc at 0'. With centre 0 and the existing radius, plus the new, strike an arc to cut the first at 0'. The new circle can now be drawn from 0'.

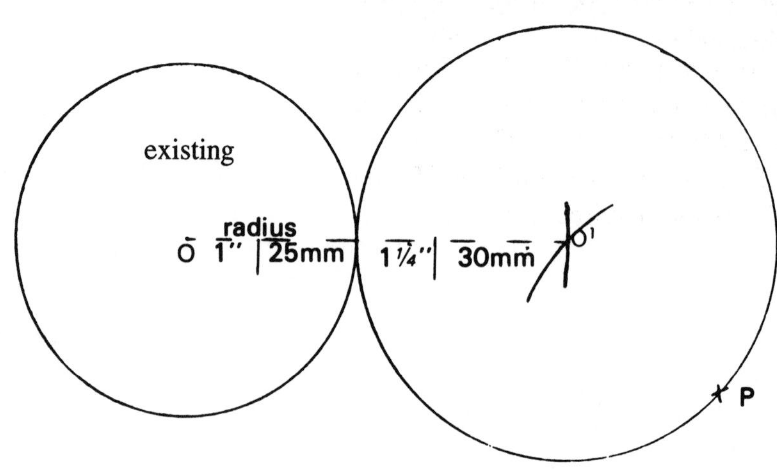

8. To draw a circle to touch an existing circle and a point C on a given line:

Erect perpendiculars to given line AB at point C and through centre 0 cutting the circle at D. Join D to C, cutting the circumference at E. Draw a line from 0 through E to cut the first perpendicular at F, which is the centre for the required circle.

Simulate existing ʘ and point C from this diagram.

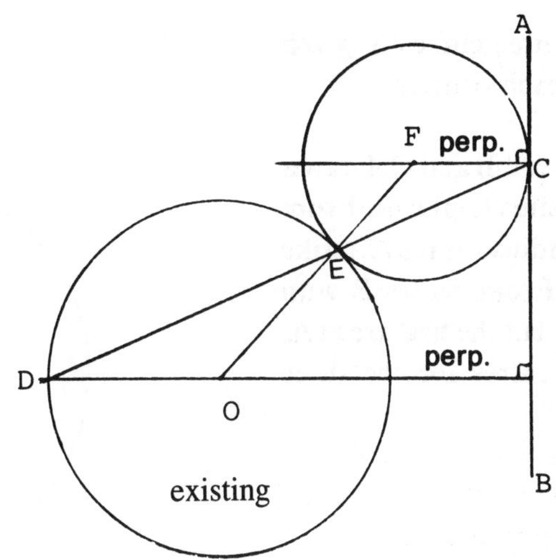

9. To draw a circle through an established point and tangential to an existing line at a given point:

Let P be the established point and C the given point of tangency. Connect point P to C. Erect a perpendicular to line AB at C and transfer the angle DCP to make EPC. The intersection O of EP and DC is the required centre.

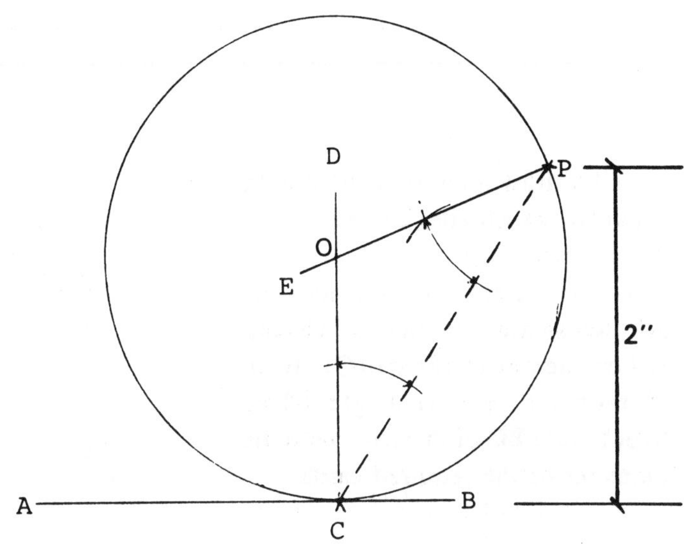

AB-P=2"	50mm
BC=¾"	20mm

10. To draw three circles of given radii to touch each other:

Let the circles be A, B and C. Draw a line equal to radius C plus B. From centre C with radius C plus A, strike an arc above. From centre B with radius B plus A, cut the first arc at A. ABC are the centres for the three circles.

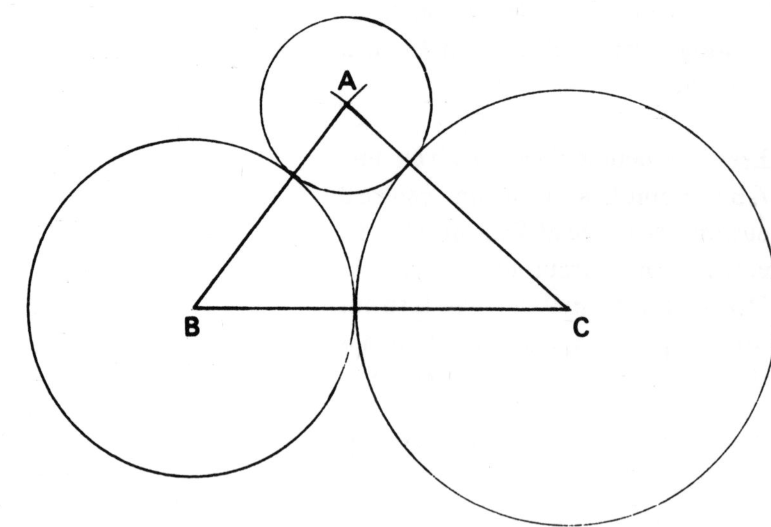

radii	A=½"	12mm
	B=1"	25mm
	C=1¼"	30mm

11. To draw a circle with an area equal to two smaller circles:

Construct a right angle and mark off the two smaller diameters, AB and AC on each side. Join B to C to form a right angle triangle. The hypotenuse BC of this triangle is the diameter of the required circle.

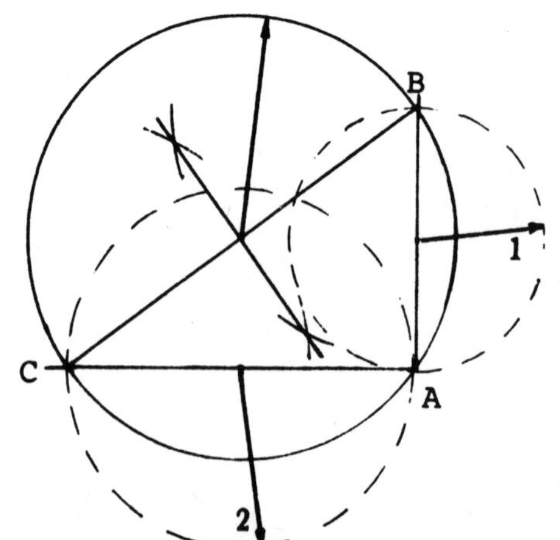

given radii 1 2

12. To inscribe a circle within a triangle:

Let ABC be an isosceles triangle. Bisect its angles. The intersection of the bisectors 0 is the centre. Set the compass to scribe a circle tangential to all sides.

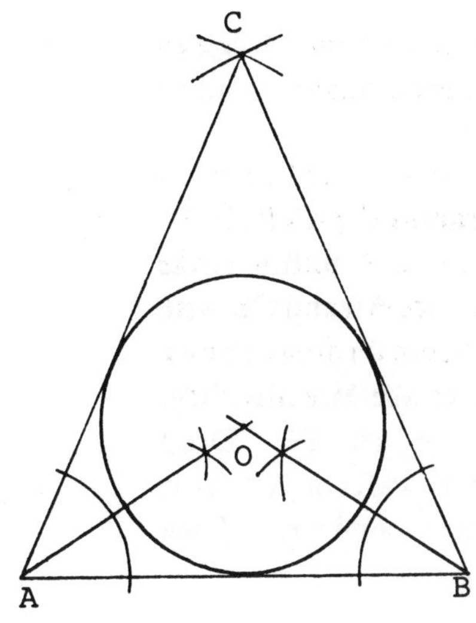

AB=2¼"	55mm
BC=2¾"	70mm

13. To circumscribe a circle about a triangle:

Let ABC be an isosceles triangle. Bisect its sides. The intersection of the bisectors 0 is the centre. Set the compass to a radius which will pass through all points of the triangle.

NOTE: Problems 12 and 13 will work for all triangles.

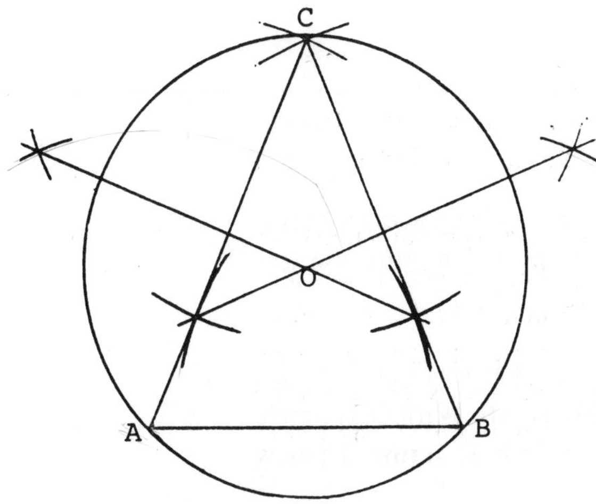

AB=1¾"	45mm
BC=2¼"	55mm

14. To draw a series of circles touching each other and tangent to two inclined lines:

Bisect the inclination of AB and CD. From point 0, draw the first circle tangent to the lines AB and CD with EO being its radius. From F, draw a perpendicular to the bisection line. From G, scribe arc FH. Drop perpendicular HO^{1} from AB. 0^{1} is the centre of the second circle. Then continue.

For practice:
∠ and radius EO may be picked up from this diagram.

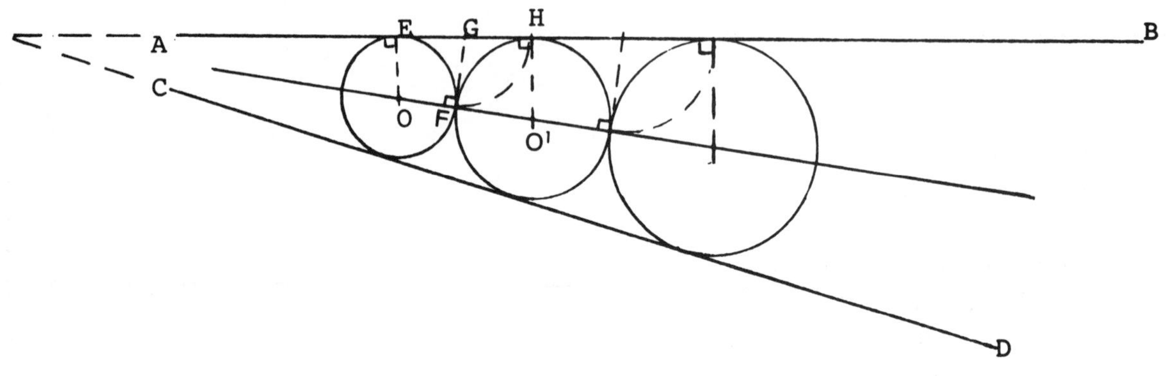

15. To draw a circle to enclose two existing circles with all three centres on the same line:

Draw a line AB through the two centres. Bisect AB. With O as centre and AO as radius, scribe the new circle.

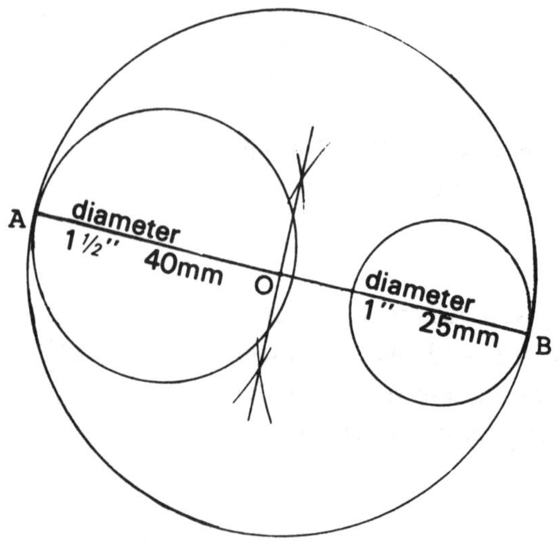

For practice:
Draw two centres 2"/50mm apart.

16. To draw a circle of given radius to enclose two existing circles:

Let the circles be A, B, and C. From centre 0 and radius of the new circle C minus radius of A, strike an arc at 0². From centre 0¹ and radius of C minus B, make another arc to cut the first arc at 0², thus locating the centre of the new circle.

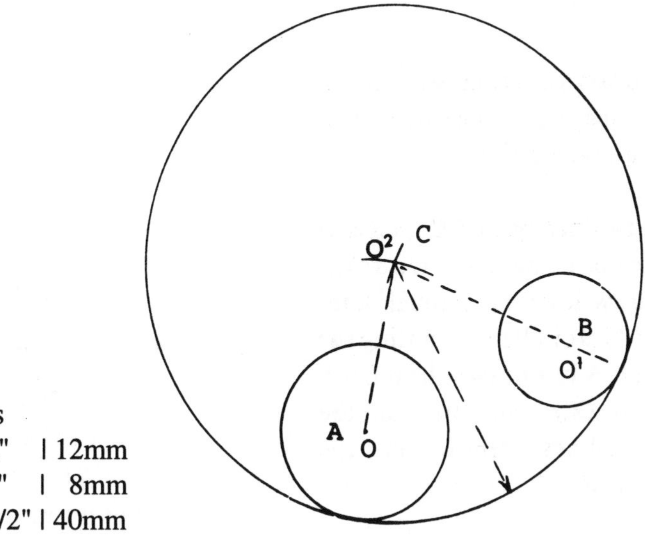

17. To draw an arc of given radius tangential to a right angle:

Set off the radius from A to B and A to C. With B and C as centres and the given radius, strike arcs at O to locate the required centre.

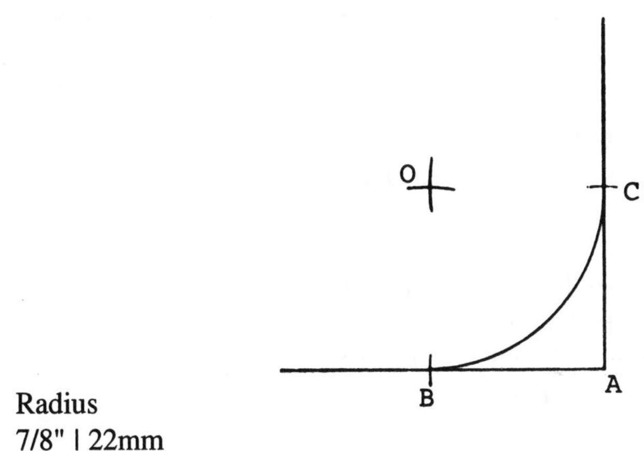

18. To draw an arc of given radius tangential to a given angle (acute or obtuse):

Using the given radius, draw arcs from lines AB and AC to form parallels. The intersection 0 of the parallels is the required centre.

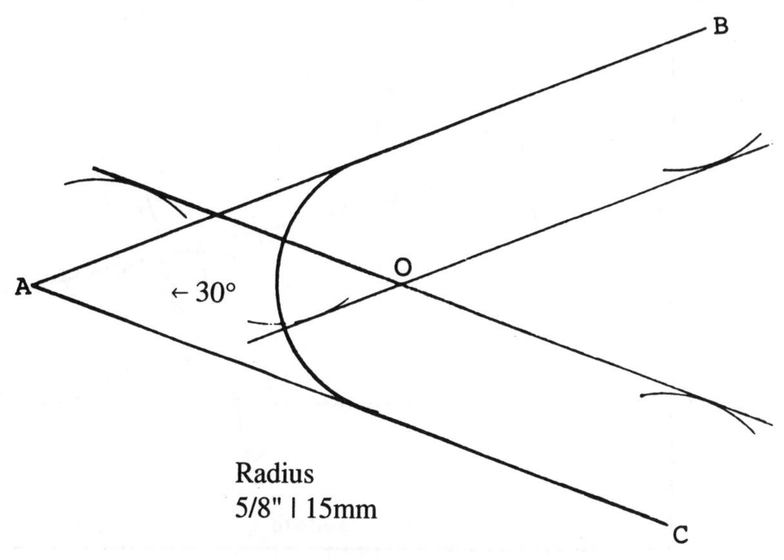

Circles

19. To draw a circle tangent to two inclined lines and also to pass through a given point P:

Bisect the given angle BAC. Select O anywhere on the bisector. With O as centre, draw a circle to touch lines AB and AC. Join P to A, cutting the circle at D. A line drawn through P, parallel to OD, will position the required circles centre on the bisector at 0'.

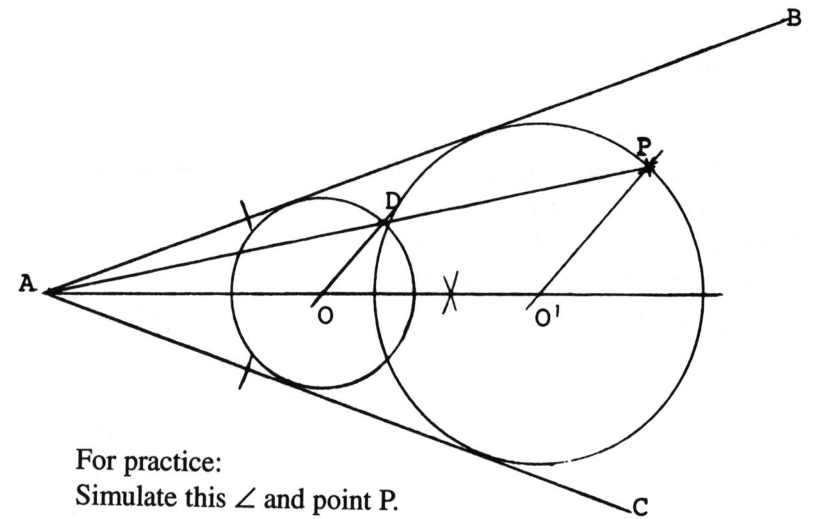

For practice:
Simulate this ∠ and point P.

20. Two pipes laid side by side of different diameters and tangential to a level base:

To layout saddles to support said pipes locate points of tangency, (T). The saddle is of 3" thick wood (fir) allowing 6" around the pipes.

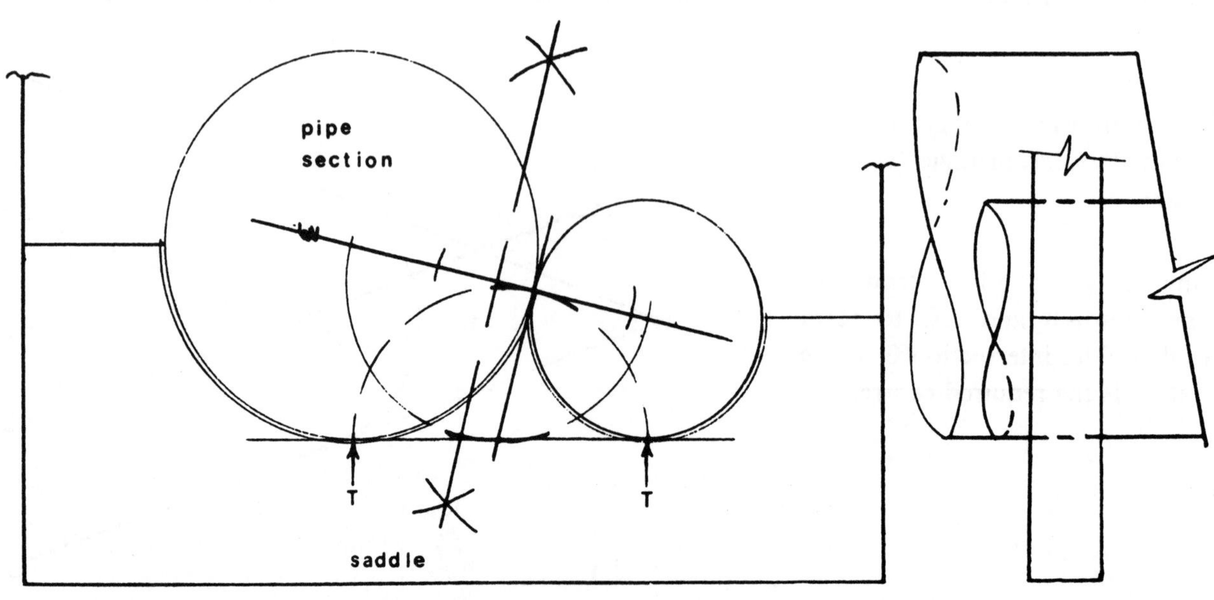

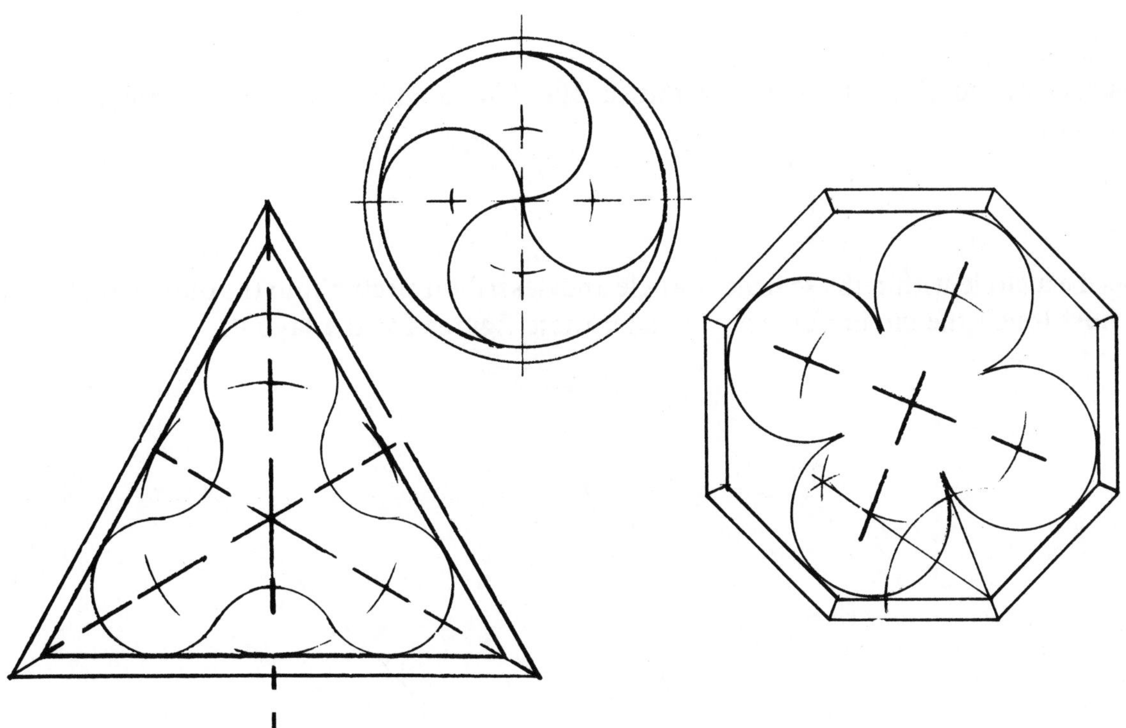

EXAMPLE OF TRACERY WORK

Discovery Exercises

1. Construct an equilateral triangle within a 2¾" (70) diameter circle. The points must touch the circumference.

2. Inscribe a circle within the scalene triangle and describe a circle about the obtuse angled triangle. The points must touch the circumference. Transfer these figures for the layout.

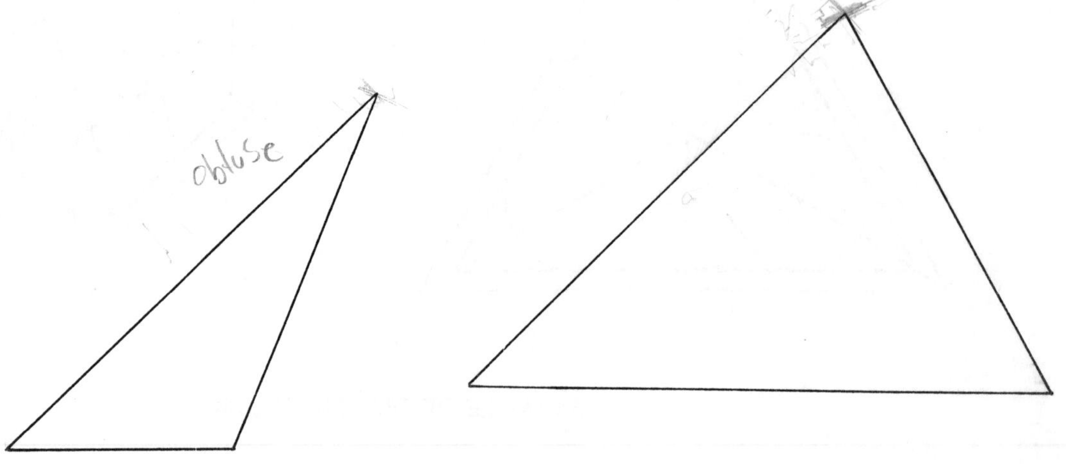

3. In a 3" (76) square, place four equal circles each touching one side of the square and two circles.

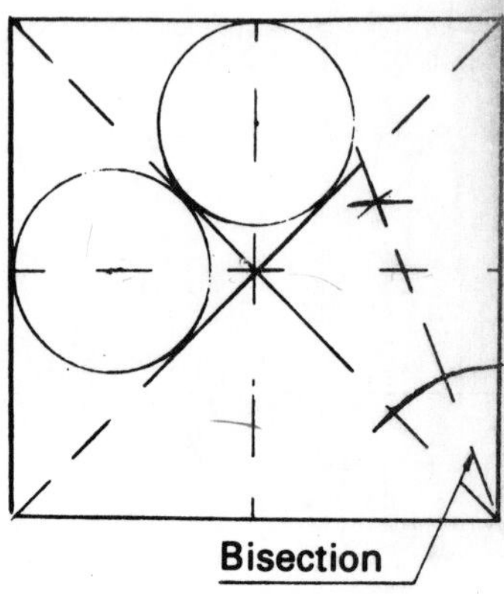

4. In a regular hexagon constructed within a 4" (100) diameter circle, place six equal circles, each touching two sides of the polygon and two other circles.

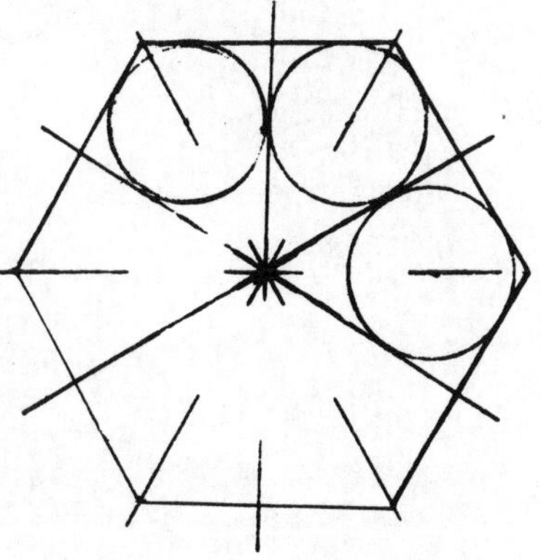

5. A round dias has to have a square built upon it. The corners of the square must be 3" (76) in from the circumference. Draw a line diagram plan of this layout and state the length of the square side. Scale 1" to 1' (1:10).

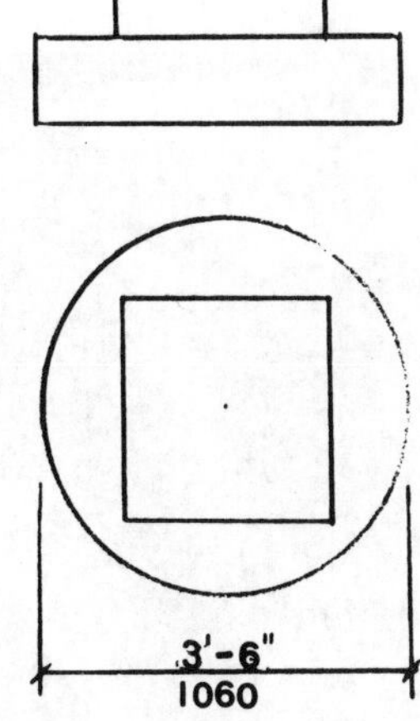

6. Layout a circle to pass through 3 given points.

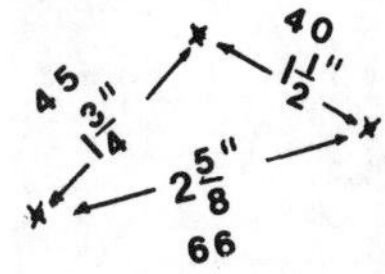

7. Layout three circles touching each other with diameters of 1½" (38), 2¼" (57), and 2¾" (73).

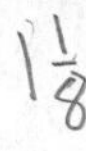

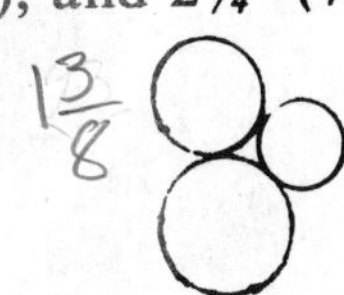

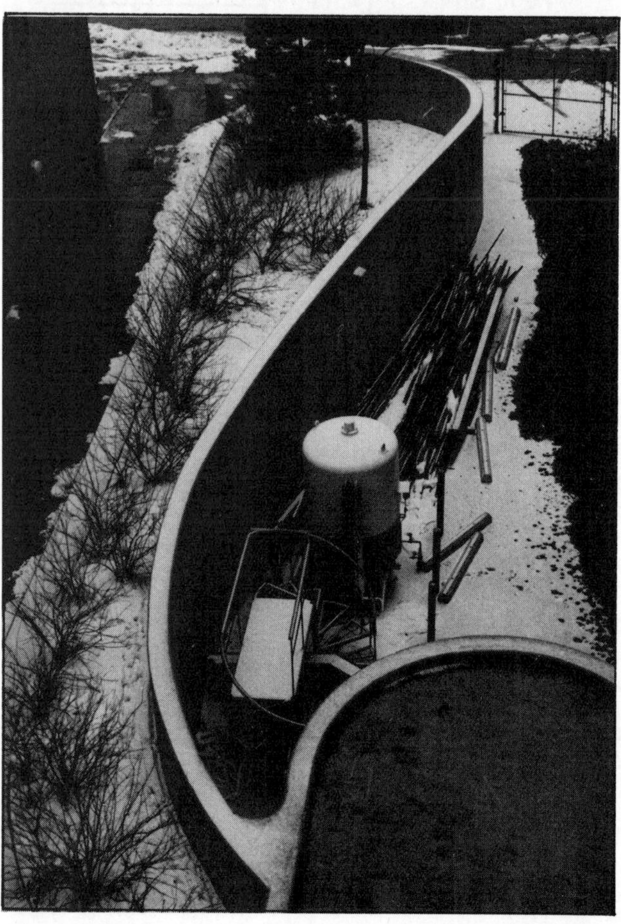

Unit 5
Ellipses

1. To draw a true ellipse by means of a wood or paper trammel: A quick, simple and practical method.

First layout the axes, AB and CD; then on the edge of a sheet of paper, or a ½" x 1" rod for larger sizes, mark off half the major axis and half the minor axis from one end. Let the measured edge lie along the minor axis with 0^1 at C. Move 0^1 to the right allowing C^1 to traverse the major axis and A^1, the minor axis. Mark many points at 0^1. Do each quarter then complete a freehand curve.

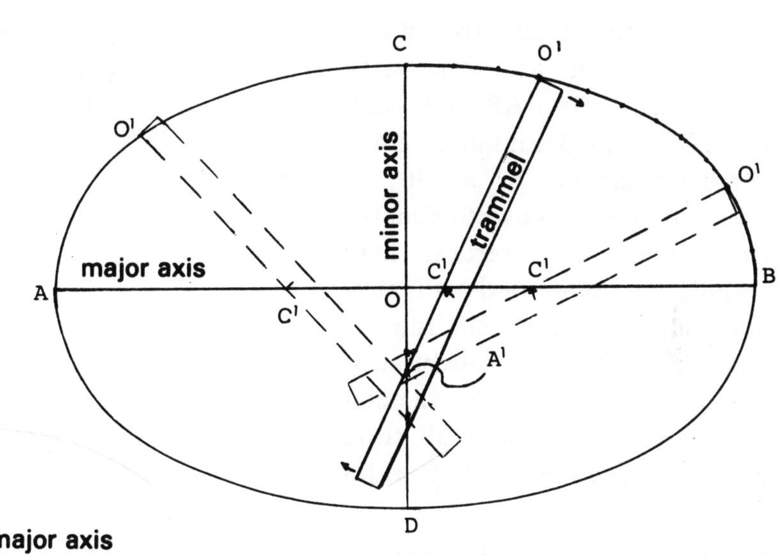

AB major axis
CD minor axis

2. Another mechanical method is with pins and string:

First layout the axes. Then find the focus points by striking arcs at F^1 and F^2 using AO as a radius from C. Secure pins at F^1 and F^2 and C. A string is now tied at F^1, stretched taut around C and tied at F^2. Remove pin C and insert a pencil. Keeping string taut, traverse to A and B.

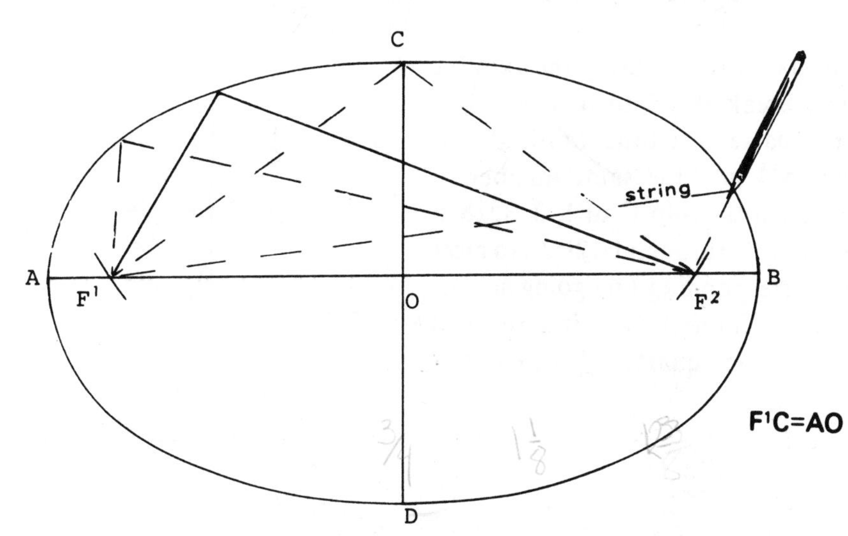

3. It can now be seen that two dimensions are usually needed to construct an ellipse. The two circle method is a geometrical construction:

Using half the minor axis and half the major axis as radii, make concentric circles AB and CD. Divide each quadrant into an equal number of sectors — 60° and 30° are used here. Where the sectors cut the circle AB, draw vertical lines (parallel to minor axis) to meet horizontal lines drawn out from the inner circle. Each of these meeting points 1, 2, 3, 4, 5, 6, 7, 8, is a point on the ellipse. Connect all these points along with AB and CD with a neat freehand curve.

NOTE: the more points found the more accurate the ellipse.

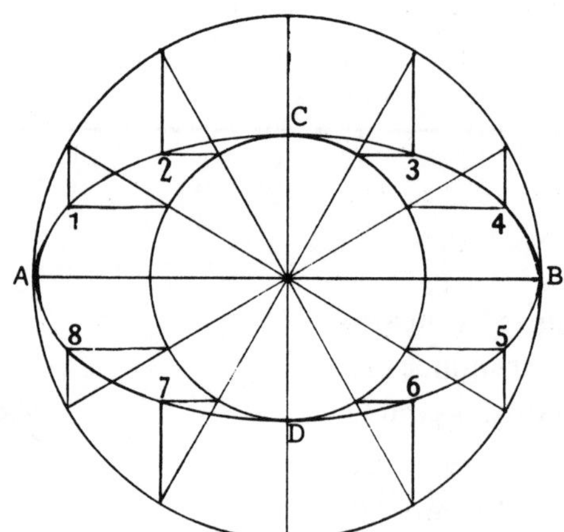

AB=3" | 80mm
CD=1¾" | 45mm

4. The rectangle method:

Form a rectangle using the axes for size; break this into quarters. Use one quarter at a time. Mark off AE and AO into the same number of equal parts. Join C to 1, 2, 3, then produce D through 1', 2', 3', to meet its corresponding line going across, 1' at 1. Connect these intersections to form one quarter of the ellipse.

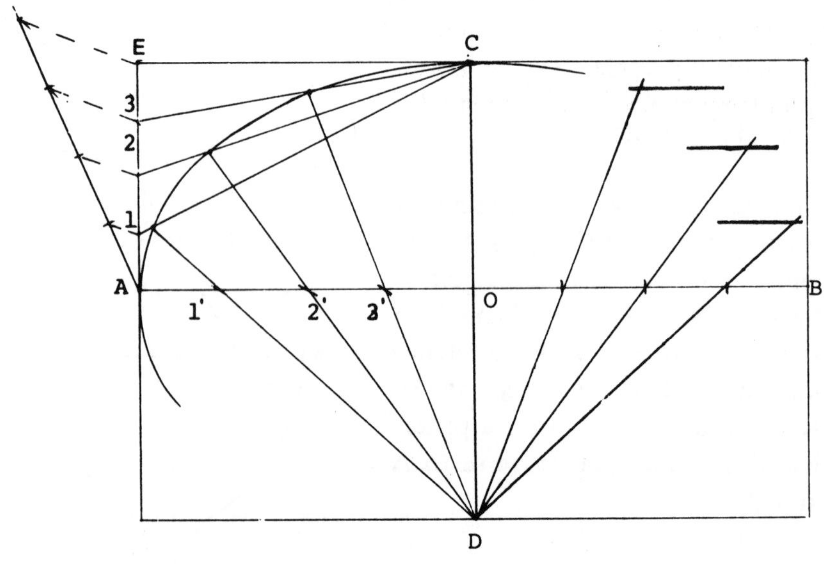

AB=5" | 130mm
CD=3" | 75mm

5. The arc method. With or without a minor axis given:

Layout the axes and mark off the focii. Divide the distance between focii into equal spaces (an odd number of spaces is best). With radius A1 and centre F^1, strike an arc at D. With radius B1 and centre F^2, cut the arc at D. Continue with radius A2 and B2 and centres F^1 and F^2 to make E and so on. Complete the elliptical curve through the intersecting arcs.

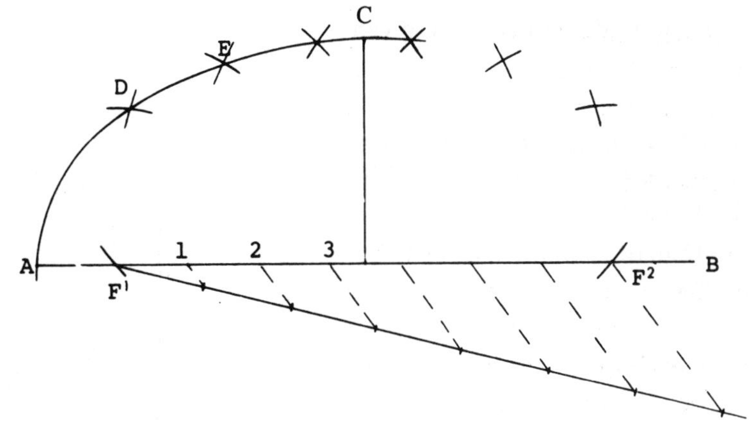

AB = 4-1/4" | 110mm
C to AB = 1-1/2" | 40mm

6. To draw a tangent and normal to an ellipse:

Extend the focii through the given point P and bisect angle FPF^1; this gives a normal. A tangent is formed by bisecting angle EPF^1. A normal and tangent are at 90° to each other.

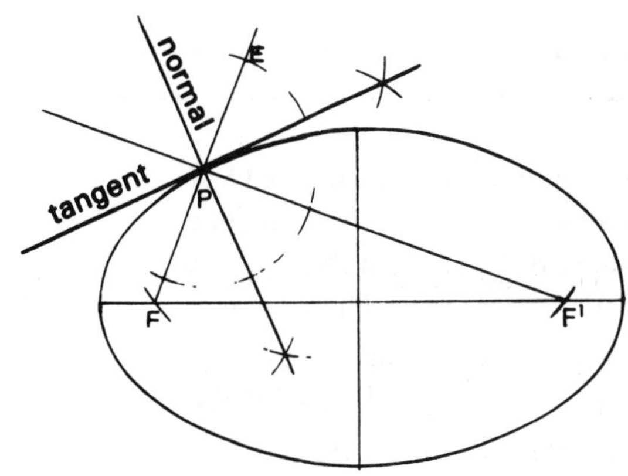

7. Finding the centre of a given ellipse:

Scribe a rectangle about the ellipse at any reasonable position. The desired centre O is the intersection of diagonals drawn through the rectangle.

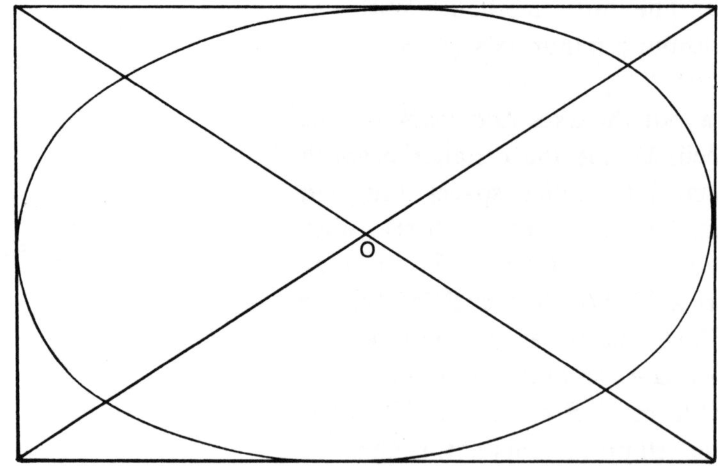

8. Finding the axes of a given ellipse:

Scribe a circle from the centre O of the ellipse to cut the curve in four places A, B, C, D. Join these points to form a rectangle. The axes can now be drawn through the centre, parallel to the sides of the rectangle.

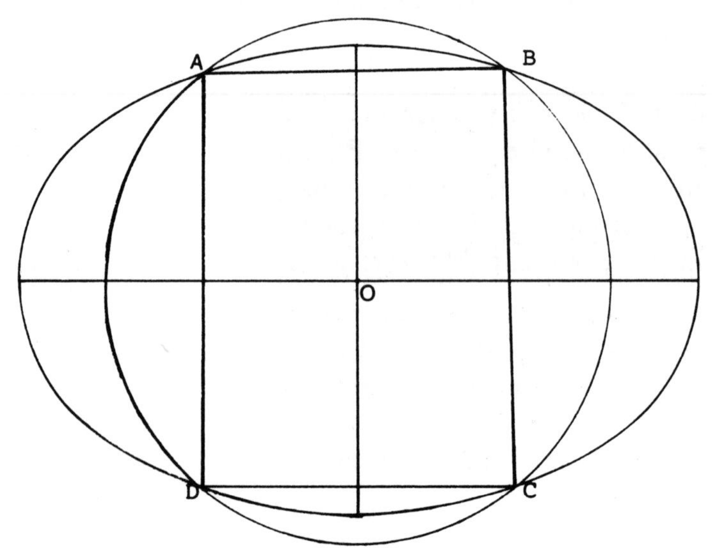

For building purposes, approximate or false ellipses (see Unit 9) are sometimes more economical to use. Though there are many varied ways of layout, just a few are discussed here.

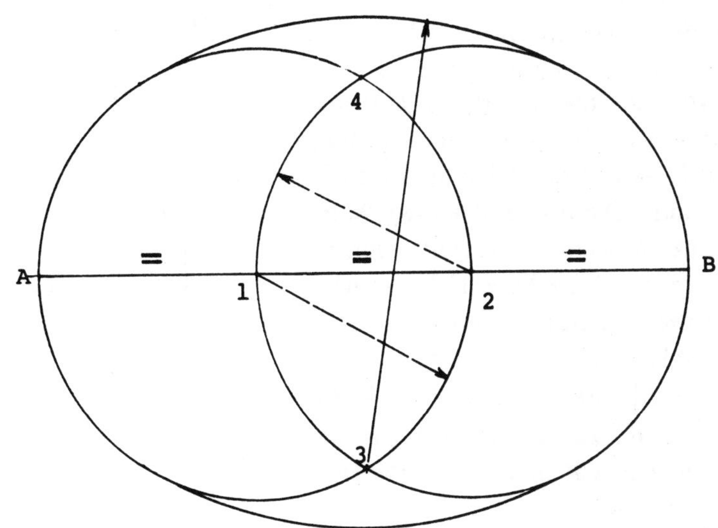

9. Given a major axes, but not a minor:

Divide the major in three equal parts. Using A1 as radius and 1 and 2 as centres, scribe two circles, the intersection of which forms centres 3 and 4. Use radii A2 from centres 3 and 4 to complete the figure.

10. Another method, which usually results in a shorter minor axis-still unknown:

Mark 1/6 of span from A and B for centres 1 and 2. Using these points and radius 1-2, make intersecting arcs at 3. Using A2 as a radius from centre 3, make an arc to blend with circles 1 and 2. Find centre 4 opposite to 3 and complete the figure.

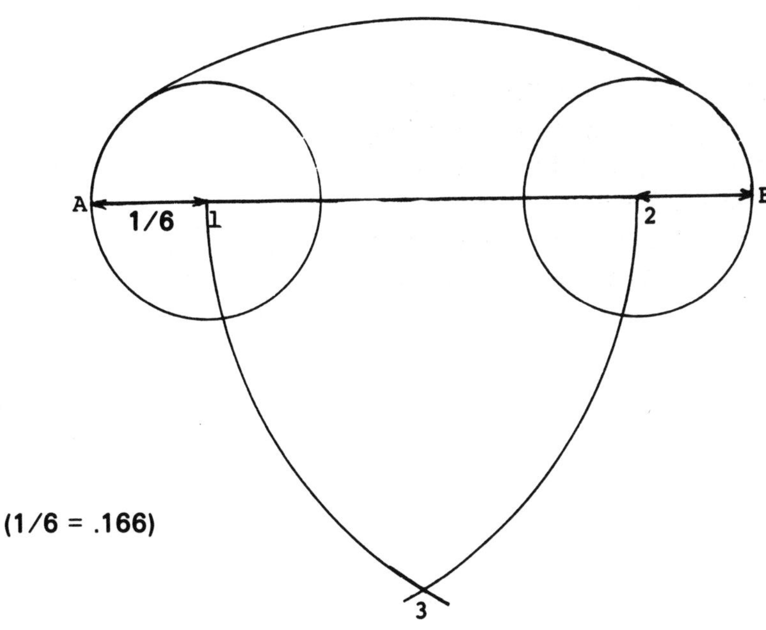

(1/6 = .166)

11. The three centred method is used when both a major and minor axes are given for an approximate ellipse:

Layout the major AB and half the minor CD. Using AD as a radius and D as centre, scribe an arc to E (extend CD above and below). With CE as radius and C as centre, scribe an arc to cut line AC at F. Bisect AF. Let the bisection cut line CD at G. Use AH as radius and centres H and I to scribe arcs to J and K. With radius GC and centre G, scribe an arc which will blend with the two smaller ones.

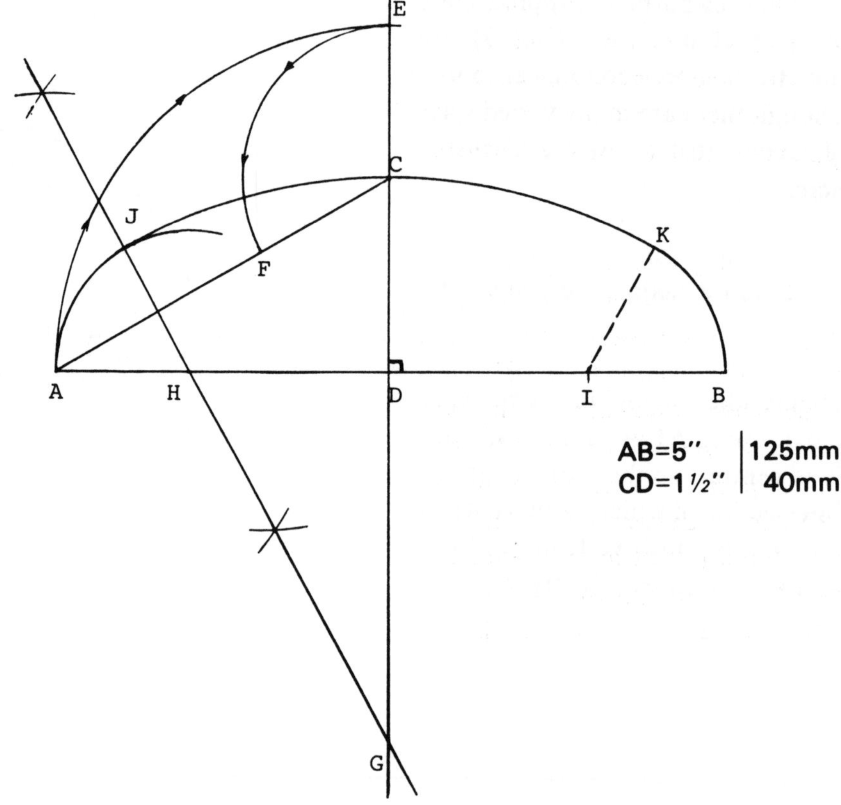

12. As explained in definitions, a true ellipse is found within a cone or cylinder. This does not mean to say that a cut through a hollow cone or cylinder would reveal parallel ellipses:

To exemplify this, a right angle cut AA is made through a cylinder. This in plan gives parallel (concentric) circles. The inclined cut BB (any angle) when viewed at right angles to the cut, gives similar ellipses BB and CC. A parallel to an ellipse such as DD can be drawn with a series of arcs E and joined by freehand. The parallel is an approximate ellipse sometimes called a pseudo ellipse.

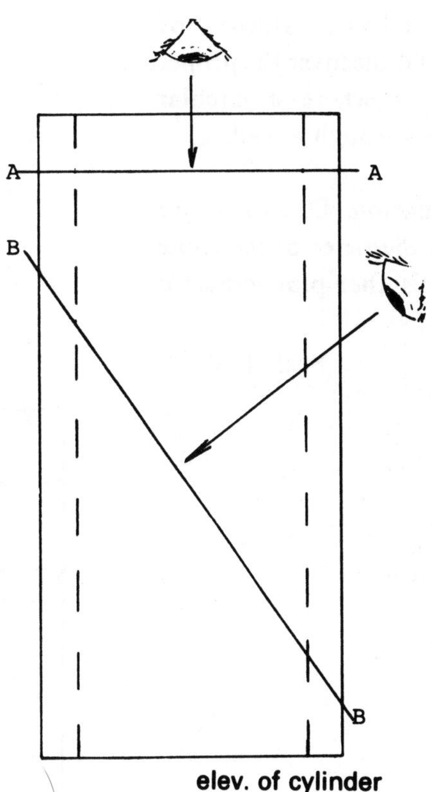

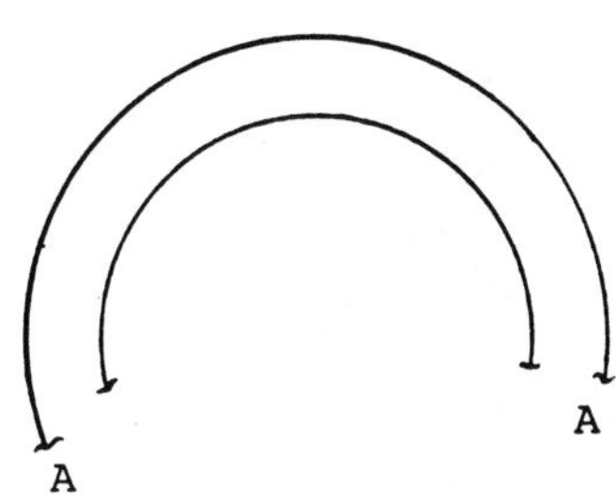

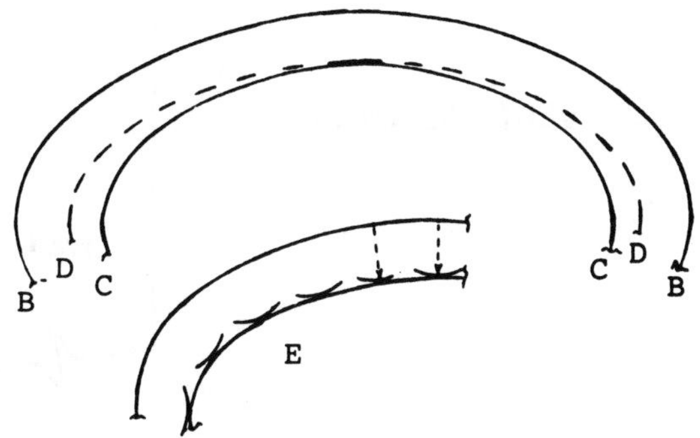

13. Finding the shape of intersection: To discover the proper elliptical shape where a circular chute inclines through a wall.

Use the dimension, D, as a major axis, and the diameter of the chute as a minor axis, then plot points for a true ellipse.

Scale 1:10

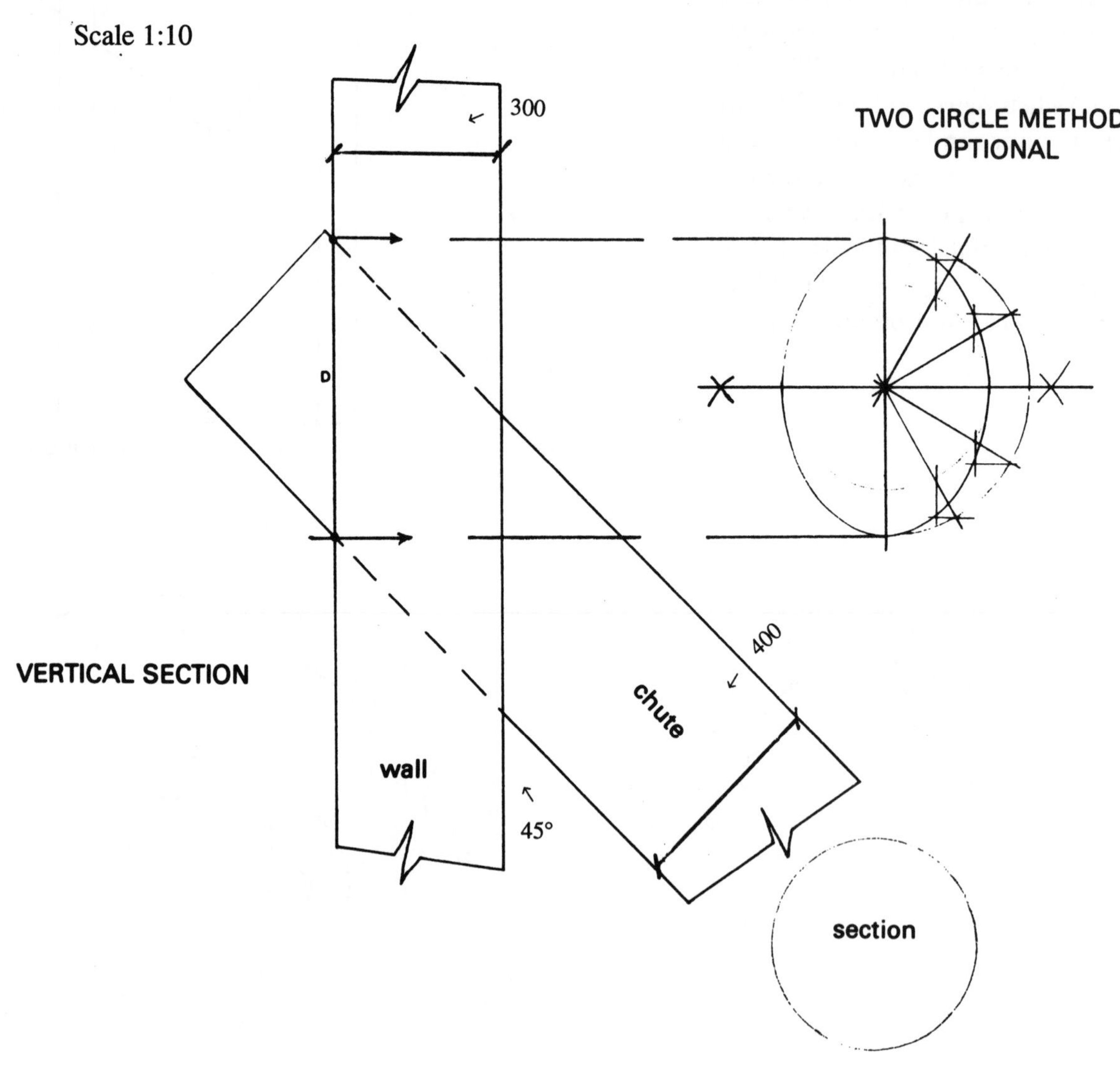

Discovery Exercises

1. Choose any recognized geometrical method to construct an ellipse having a major axis of 3¾" (95) and a minor axis of 2½" (64).

2. A board fixed at an angle of 30° must have a hole cut through it to allow a vertical 3" (75) O.D. pipe to fit snugly into it. Draw the true shape of the hole and state the length of both axes.

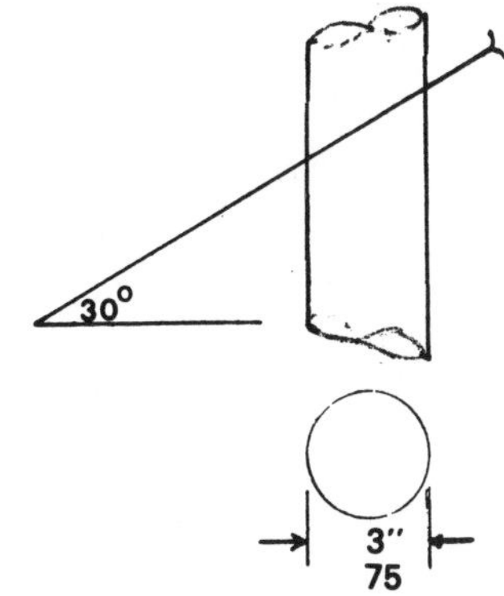

3. Transfer this figure and:
i) At point a, draw a tangent.
ii) At point b, draw a normal.

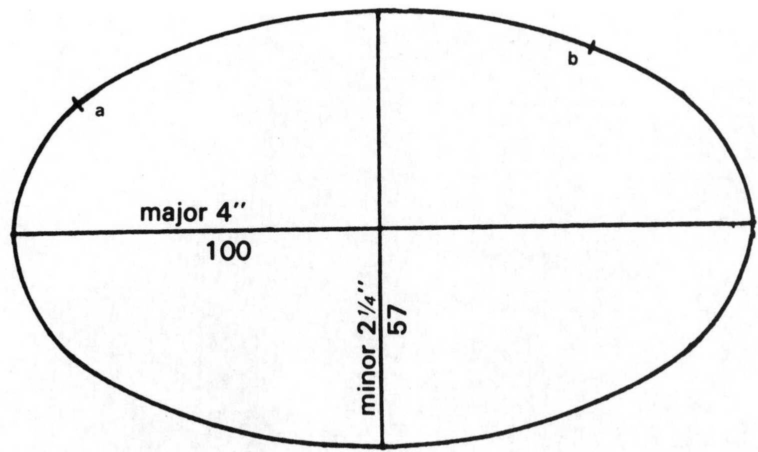

4. Illustrate by sketch how the pin and string method works to layout an ellipse. Label all lines.

Unit
Polygons

REGULAR POLYGONS

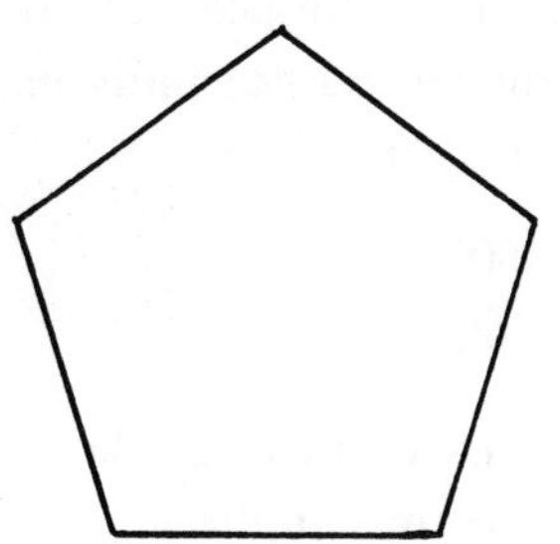

Pentagon

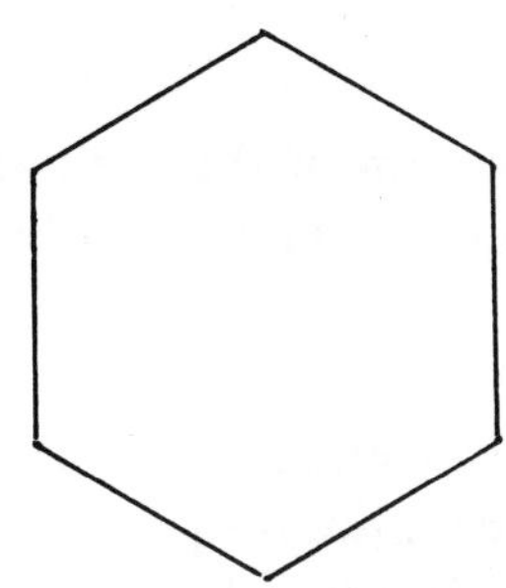

Hexagon

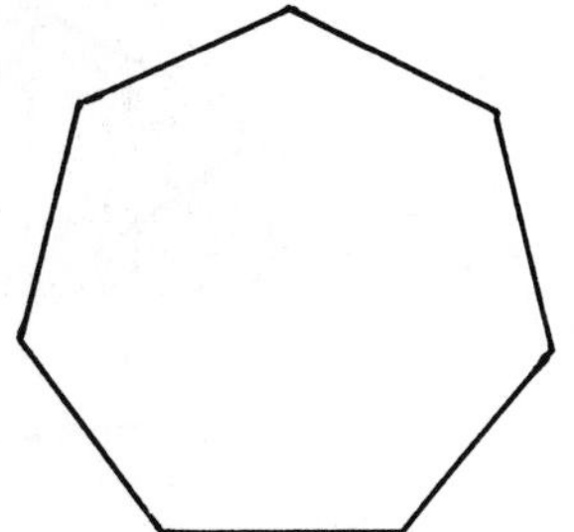

Heptagon

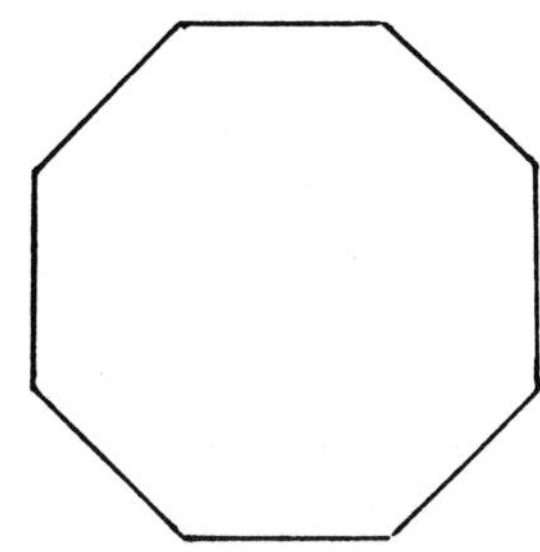

Octagon

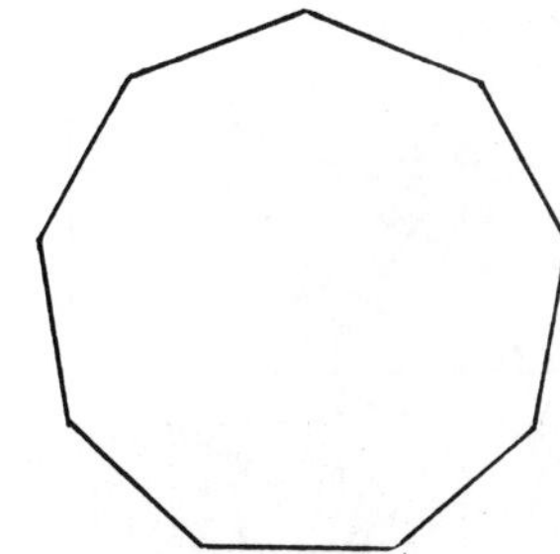

Nonagon

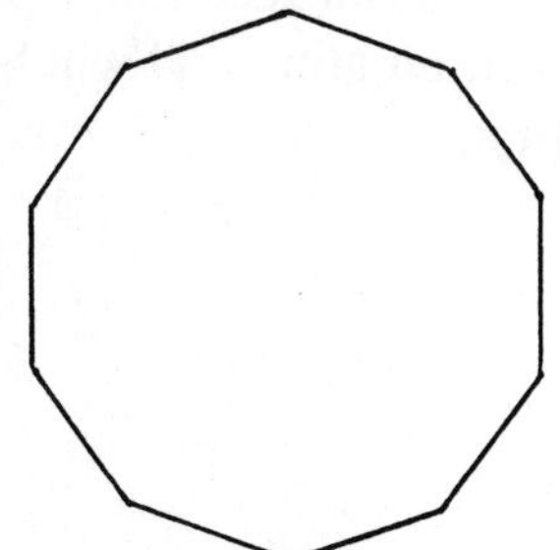

Decagon

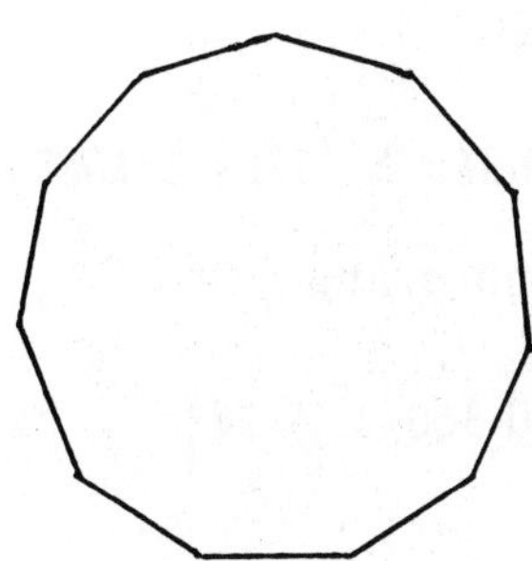

Undecagon

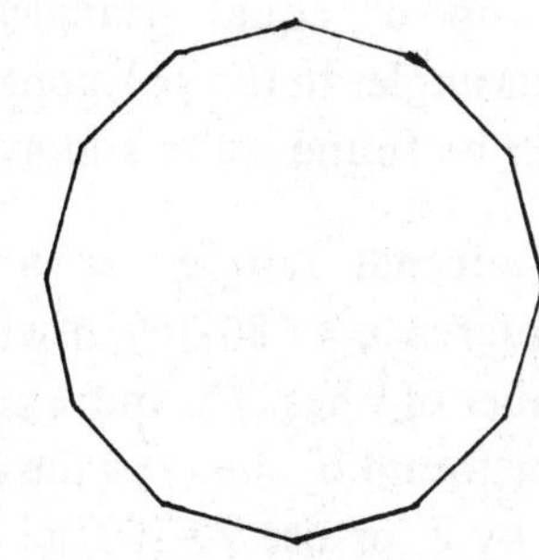

Dhodecagon

Polygons

The geometrical properties of polygons are:

1. Polygons are formed by a series of triangles.

2. The sum of the external angles x=360°.

3. The sum of the apex angles Θ (polygen centre) of each triangle = 360°.

4. The external angle x of a regular polygon is always equal to the centre angle Θ it subtends.

5. Polygon measurements may be given by: the length of a side, the distance across point to point, side to side, or side to point. The method of layout is then determined by what information is known.

NOTE: The angles of irregular polygons would normally be solved by trigonometry.

5x — 360°
5Θ — 360°

Θ — Apex (centre) angle
X — External angle

A — Internal angle
B — Mitre angle

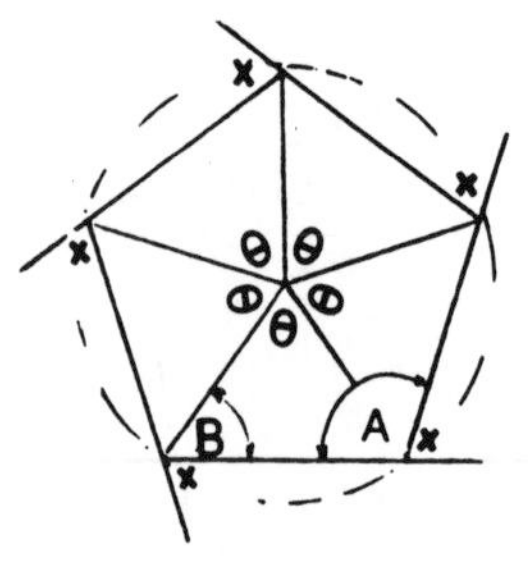

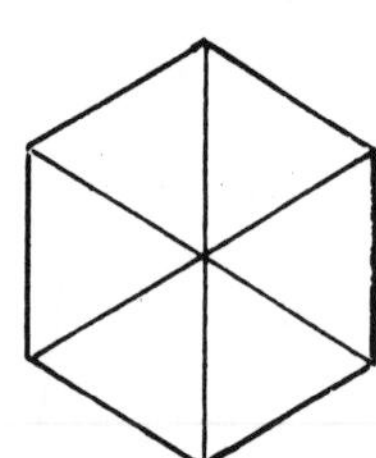

Regular

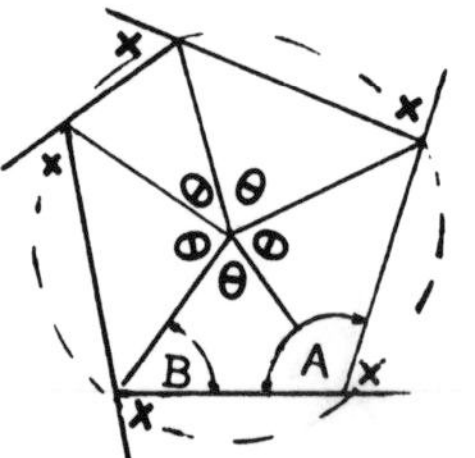

Irregular

6. With regular polygons being made up of equal triangles, the various angles in the polygons make up can be found quite simply:

The internal angle A at the circumference = 180-360/n where n = number of sides. The mitre angle B may be found by dividing the above result by 2, or use 90-180/n.

Example: The internal angle of a pentagon is:

A=180-360/5 180-72=108°

The mitre angle is:

B=(180-360/5)½=54°

or:

B=90-180/5 90-36=54°

The hexagon and octagon are probably the most common polygon shapes in building and will be dealt with first.

Hexagon

1. To inscribe a hexagon in a given circle (or measurement point to point):

Use the radius of the circle $00^{\prime}$ and step off six times around the circumference. Join points AB, CD, EF. The radius of a circle will always step off the circumference six times. The radius of a circle will always equal the length of a side of a hexagon contained within the circle. The hexagon is made up of 60° equilateral triangles.

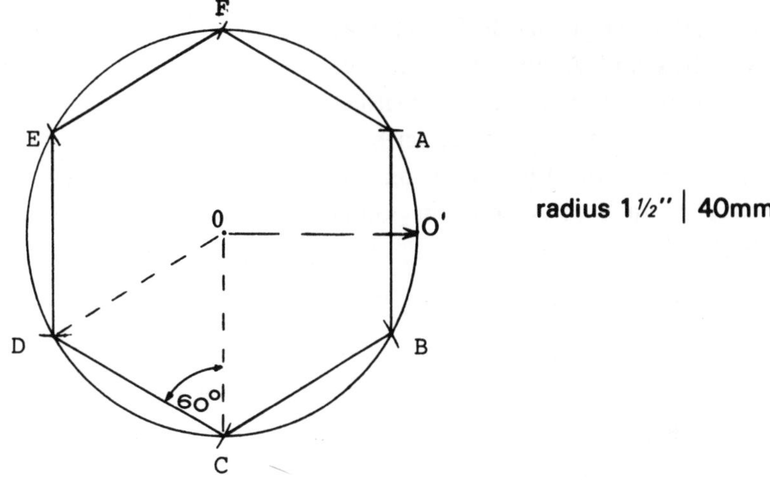

2. Given the distance across the points of a hexagon as a diameter:

Scribe a circle. Draw two diameters at 90° to each other, AB and CD. Then at 30° to the horizontal, make EF and GH through the centre. Join CFHDEG.

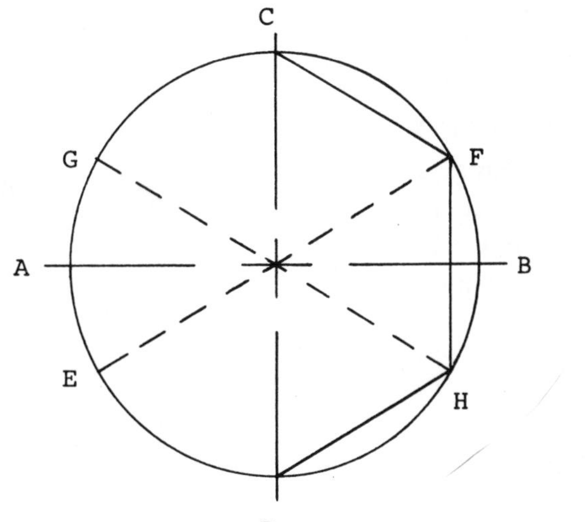

diameter 3" | 75mm

3. Given the length of one side of a hexagon:

Let AB be the given side. With AB as a radius and A and B as centres, scribe arcs to intersect at O. With O as centre and radius AB, scribe a circle. Step off the radius on the circle from B to A and join points BCDEFA.

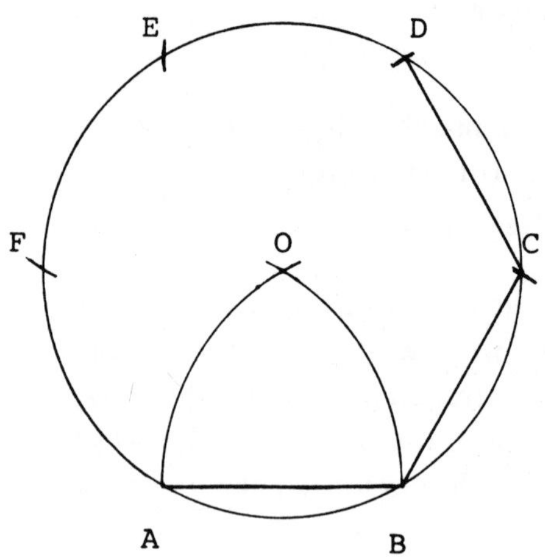

AB=1 ¼" | 30mm

4. Given the distance across from side to side of a hexagon:

Draw two parallel lines AB and CD the given distance apart. Erect a perpendicular EF and bisect it. Through this centre 0, draw 60° lines GH, IJ. Make further 60° lines LH, JL, IK, KG.

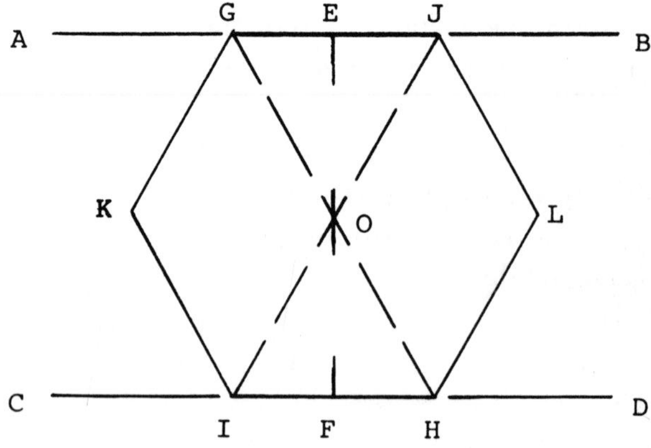

EF=3" | 75mm

5. To draw a hexagon within an equilateral triangle:

Let ABC represent the triangle and find its geometrical centre by bisecting the angles. Use centre O and circumscribe a circle to pass through points ABC. Extend the bisection lines to cut the circle at DEF. Join DEF to form an inverted triangle. The hexagon can now readily be seen in the overlap.

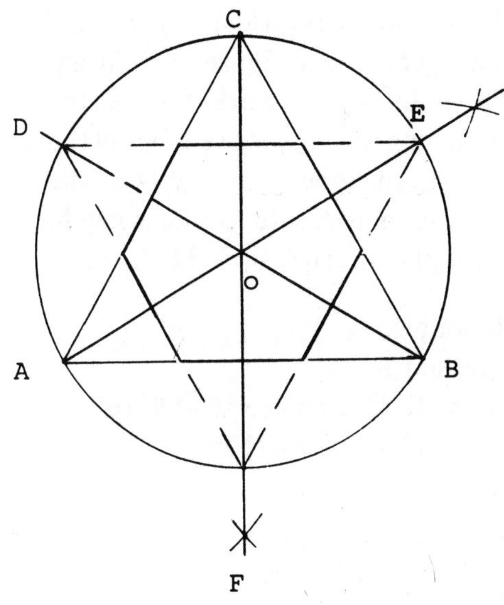

AB=3" | 75mm

Octagon

6. To inscribe an octagon in a given circle (or measurement point to point):

Draw the circle and two diameters AB and CD at 90° to each other. Bisect the right angles letting all bisection lines touch the circle. Join AGCFBHDEA to form the octagon. The octagon is made up of isosceles triangles.

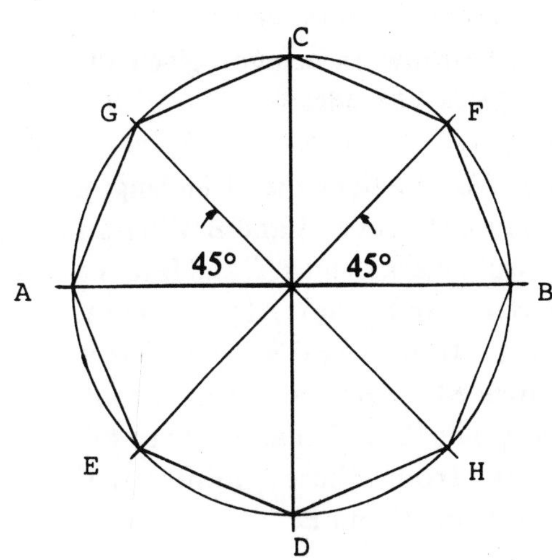

AB=3" | 75mm

The octagon is possibly the most used geometrical figure in building. Often in layout work one of the following formula is used to find the length of one side. When the distance across side to side is given, multiply by .4142 or 5/12.

Example:
Method 1
X=.4142 x 15'=6.213'=6'-2 9/16"
X=.4142 x 5m=2.0710m

or
Method 2
15' x 5/12=6.25'=6'-3"

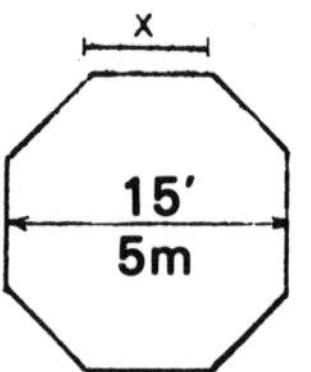

As you see the results are close. Method 2 is approximate, yet will suffice for many practical situations. Method 1 is derived from a table of natural tangents for the angle of 22½° and is considered quite accurate.

7. To draw an octagon given the length of one side:

Let AB represent the given length. Draw 45° lines at A and B with a set square; mark them at C and H to the given length. Next draw vertical lines from ABCH. 45° lines extended from A and B will determine D and G and further 45° lines from these points will determine F and E.

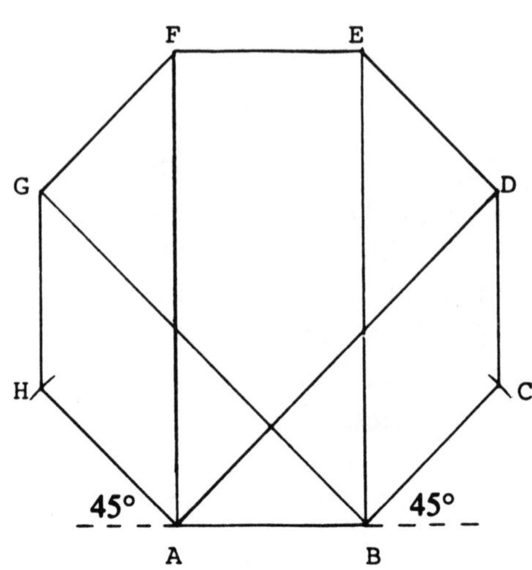

AB=1¼" | 30mm

8. To draw an octagon given the distance across side to side:

First construct a square ABCD. Draw two diagonals. With radius AO and centres ABCD, make four arcs EH, GJ, IL, KF. Join these new points on the square to form the octagon.

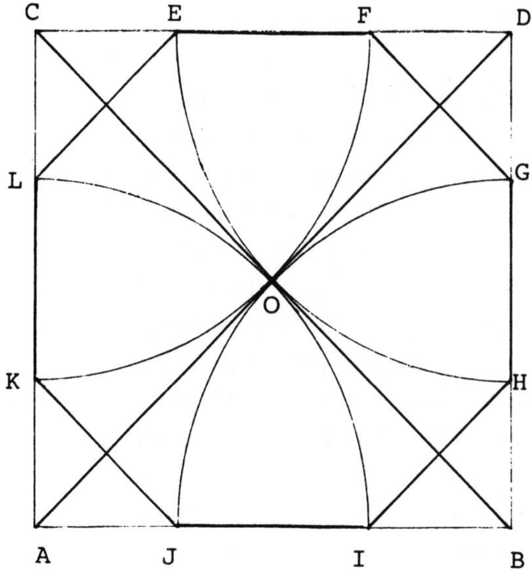

AB=3" | 75mm

9. To draw any regular polygon on its given side:

Let AB be the given side of a pentagon. Extend AB to the right. Use B as a centre and use a convenient radius to draw a semicircle. Divide this into an equal number of parts that the polygon has sides. Through division 2, extend a line BC equal in length to AB (always through 2 to find the second line's inclination). Bisect AB and BC to find 0. Using 0 as centre, scribe a circle to pass through points ABC. AB may now be stepped off about the circle for points D and E.

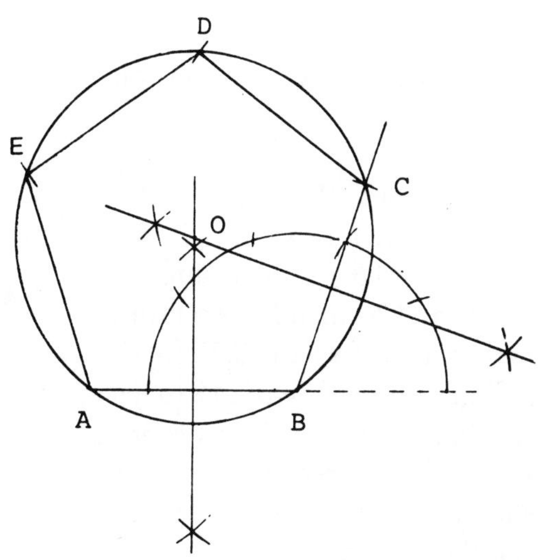

AB=1 ½" | 40mm

10. Alternate to number 9:

Let AB be the given side. Erect a perpendicular at B equal in length to AB. With AB as radius and B as centre, scribe arc AC; also join AC with a straight line. Bisect AB with an extended line. Where the bisection cuts AC it gives centres 4 and 6. Divide 4, 6 to get centre 5. The same divisions can now be carefully marked off on the bisection line for more centre points. These are centres for circles containing polygons with sides equal to the given side, which may be stepped off on the circumference.

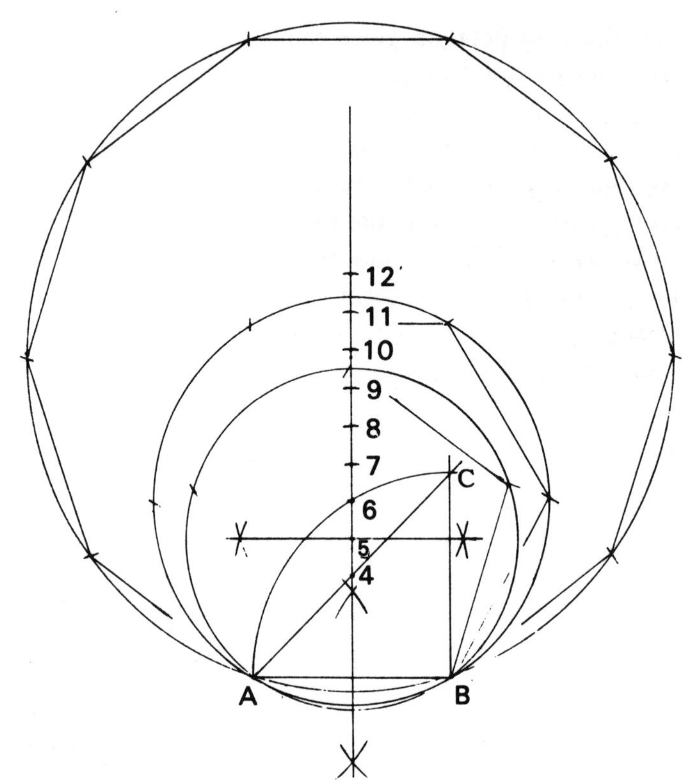

11. To inscribe a regular polygon in a given circle:

Let AB represent the diameter of a circle to receive a heptagon. Divide AB into 7 equal parts, then using AB as radius and A, then B, as centres, scribe arcs to intersect at C. From C, draw a line through division 2 to the circumference at D. AD equals one side of the required polygon, the remaining sides can now be stepped off on the circumference.

NOTE: To circumscribe a polygon, first use above method then extend radial lines through each point and draw parallel lines to the existing sides, as in Figure 11a.

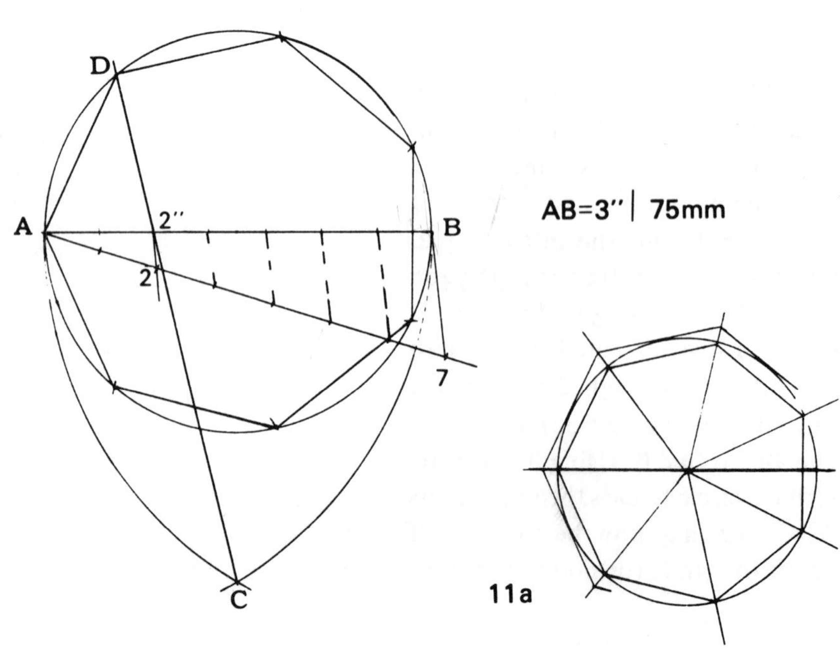

12. Finding polygon sides: To layout 3 sides of an octagon for bay window walls and footings.

Given dimension AB, layout half a square for the octagon, AD=half AB. Use centre A and radius AO to strike C. Transfer the gauge CB to the building line at both sides leaving X the octagon side. Strike X^1 to intersect at E thus giving the three sides.

Scale 1:20
AB = 3000

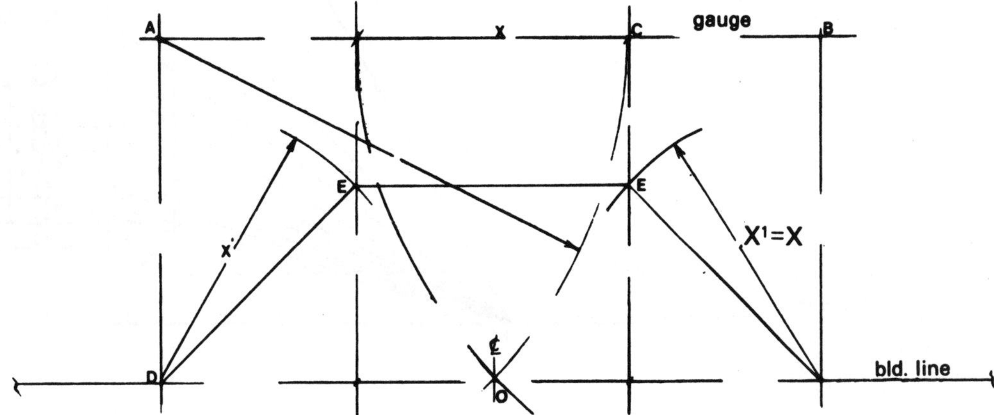

Polygons

Mitre Angles

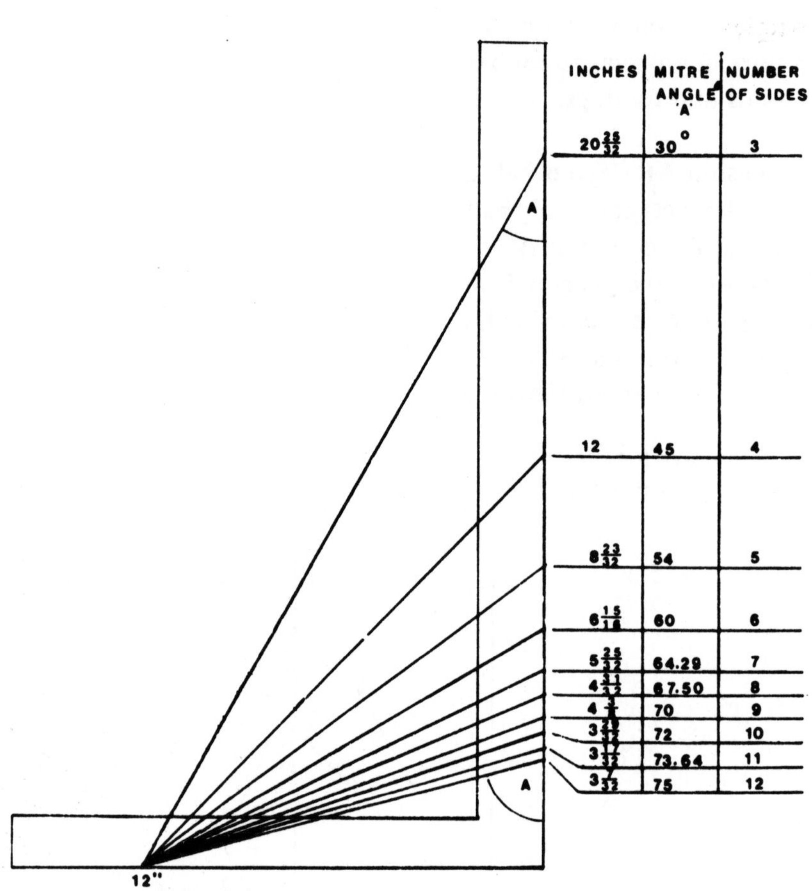

INCHES
MITRE ANGLE 'A'
NUMBER OF SIDES
20 25/32 30° 3
12 45 4
8 23/32 54 5
6 15/16 60 6
5 25/32 64.29 7
4 31/32 67.50 8
70 9
72 10
3 17/32 73.64 11
3 7/32 75 12
A
A
12"

Mitre Angles

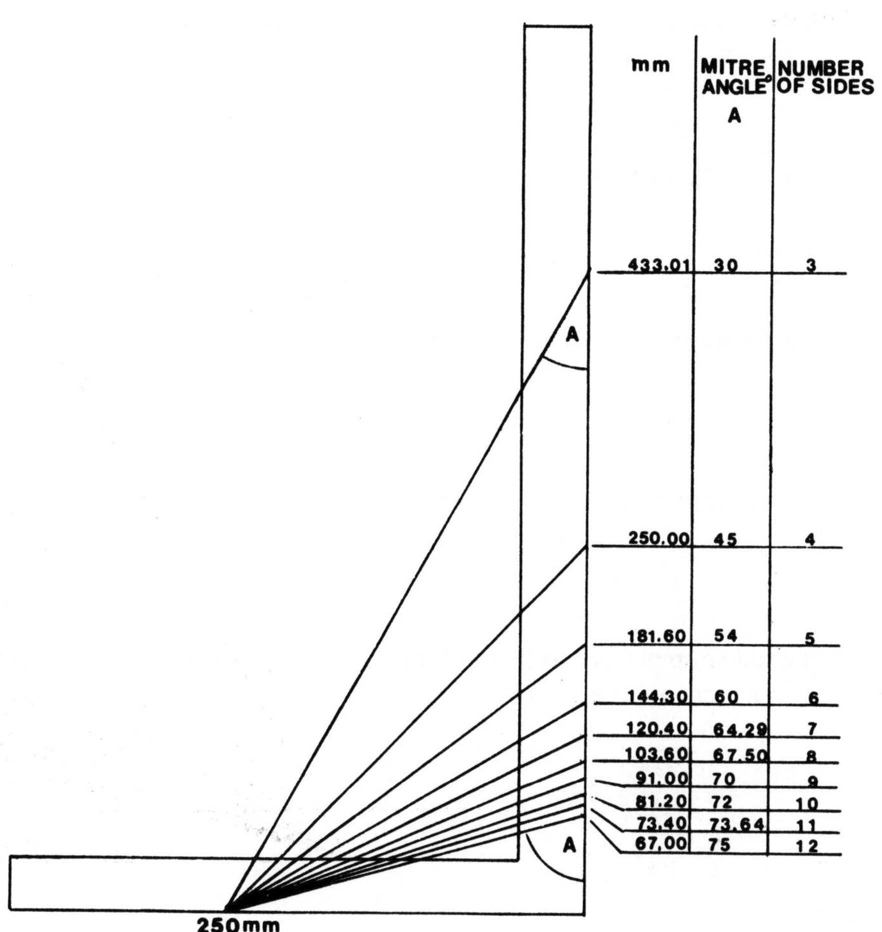

mm
MITRE ANGLE° A
NUMBER OF SIDES
433.01 30 3
250.00 45 4
181.60 54 5
144.30 60 6
120.40 64.29 7
103.60 67.50 8
91.00 70 9
81.20 72 10
73.40 73.64 11
67.00 75 12
A
A
250mm

Discovery Exercises

1. Layout an octagon with 1½" (40) sides.

2. Layout a heptagon with 1¼" (32) sides.

3. Layout a pentagon within a 4" (100) circle.

4. One hundred shapes have to be cut from ½" (12) birch. Transfer the pattern to full size for a template. The design is based on the octagon.

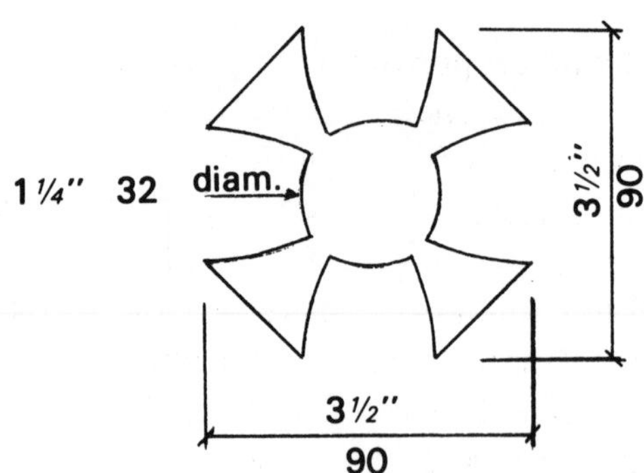

Unit 7
Ratio and Proportion

The solution to a proportioning problem may be accomplished, with reasonable accuracy, by simple straight line drawings.

A ratio is a relationship between two quantities given in any shape or form. Example: 2 is to 4 (written 2:4 or 2/4). 2:4 and 5:10 are similar insomuch as their division abilities. Therefore, it is correct to state that they are in the same ratio.

To divide proportionally means an equal or just share. In proportion problems, there is usually two ratios. Example: 2 : 4 as 5 : 10 (written 2 : 4 :: 5 : 10 or 2/4 x 5/10). The four numbers make up the terms of a proportion. The outer terms are called the extremes, the middle terms are called the means.

If one term is unknown, it is a fourth proportional problem. If two terms are unknown, it is a third, or possibly a mean proportional problem.

To solve, multiply the extremes. This will equal the product of the means. The unknown, (x), is normally positioned as the end term, except in a mean proportion.

Example: Find the fourth proportional to numbers 2, 4, 5.

2 : 4 :: 5 : x
2x=20
∴ x=20/2
∴ x=10

Example: Find the third proportional to numbers 3, 6. (Use the last term twice.)

3 : 6 :: 6 : x
3x=36
∴ x=36/3
∴ x=12

We will now proceed to accomplish similar results with line drawings and measuring the answer. The mathematical proof is shown only for comparison.

1. A line divided into equal proportions by parallel lines. (See Straight Lines, problem 8).

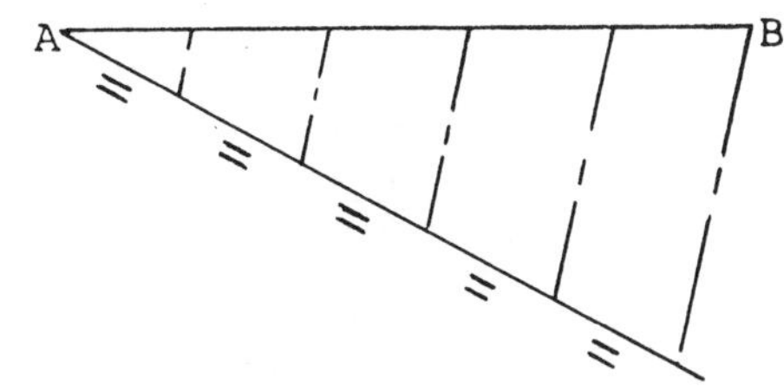

2. To divide a line AB into varied proportions:

Total the given units, 2+3+4=9. Divide line AC into 9 equal units. Join 9 to B. Parallel to this, connect 2 units to line AB, then 3 more units.

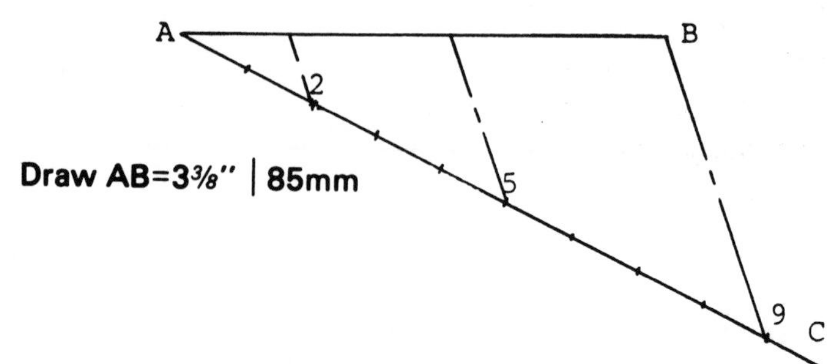

Similar triangles are proportionate.

Many tradesmen use ratio and proportion to solve various problems. A good example of its application is similar triangles. They are similar insomuch as the angles must remain the same, but the length of the sides can be increased or decreased proportionally.

3. A proportionally divided triangle:

DE is parallel to AC, thus dividing AB and CB into similar proportions.

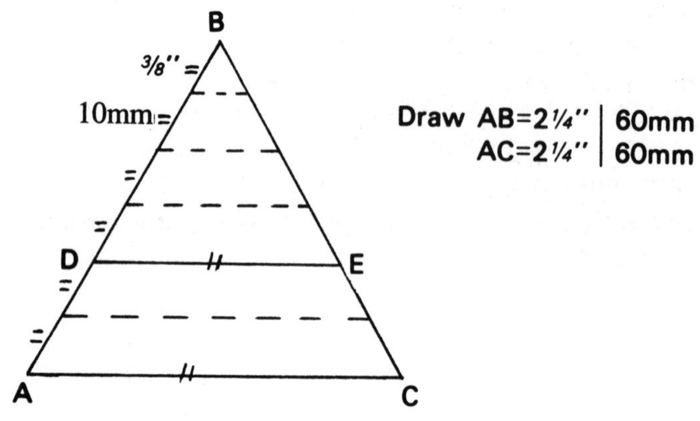

ratio=2:4
AD:AB::CE:CB
AB and CB are proportioned equally.

4. As long as a dividing line is parallel to one side of a triangle, the other two sides are divided proportionally and will have the same ratio.

Draw		
AB=2¾"	70mm
AC=2"	50mm
EC=½"	12mm

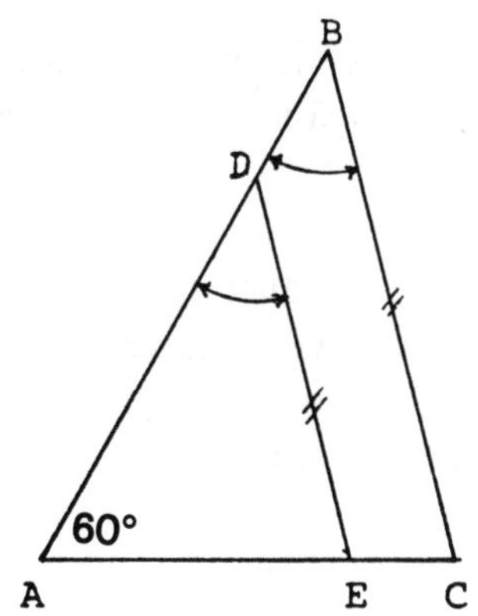

DE || BC

Proof:

AD is proportionate to AB

AE is proportionate to AC

ADE is similar to ABC.

5. To increase or decrease a triangle and maintain identical shape:

Given abc, find the lengths of a similar triangle with a base of 8 units. By drawing, transfer the angle cab with a compass. By math, the solution is shown.

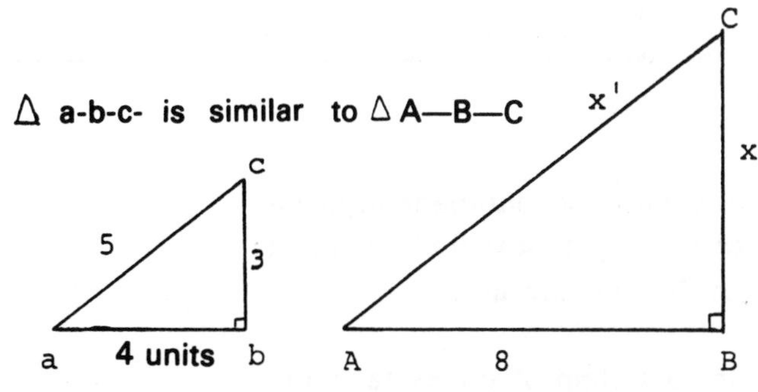

Proof: ab : AB :: bc : BC

4 : 8 :: 3 : X

4x=24

x=24/4

x=6=CB

Make 1 unit = ¼"|10mm

also 4:5::8:X¹

4X =40

∴ X¹=40/4

∴ X¹=10=AC

6. This is not a right angled triangle, but the same principles apply:

Told to reduce triangle ABC to 10 on side AB, find the length of the other two sides of the smaller, similar triangle.

NOTE: If no angles are given, BE must be calculated.

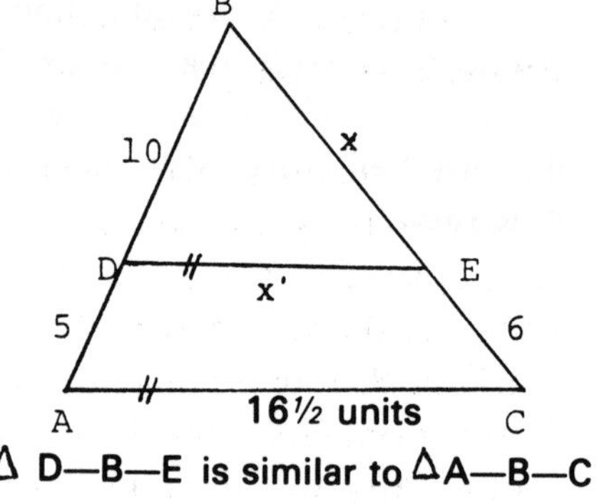

make 1 unit= ⅛" or 3mm

Proof: find for BE

5:10 :: 6:X

5X=60

∴ X=12=BE

Find for DE

$15:16\frac{1}{2}::10:X^1$

$15X^1=165$

$X^1=11=DE$

7. A typical layout diagram that a surveyor might use to find distance x when B is unreachable:

Using a system of stakes, tape and transit, x may be found.

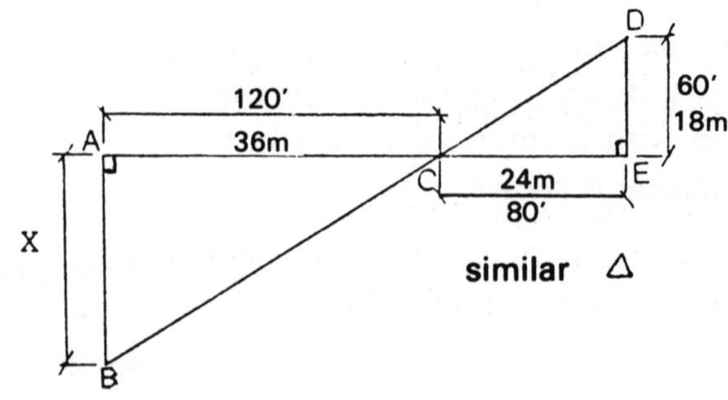

Find for AB

Proof: 80:60::120:X

80X=7200

X = 90'-0" = AB or 27m

8. To draw a fourth proportional greater than three given lines:

Let ABC represent the given lines. Draw two lines at an acute angle to each other. On the base, mark off A and C as shown. On the other line, mark off B. Join length A to length B and parallel to this draw a line from length C. The fourth proportion can now be measured as 3".

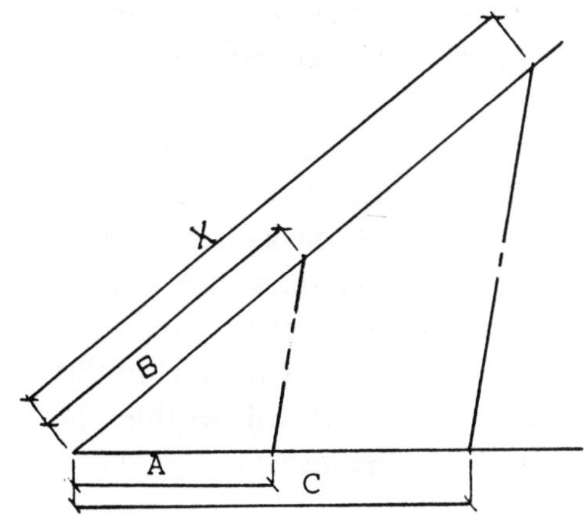

A	1"	25mm
B	1½"	40mm
C	2"	50mm

Fourth Proportional Greater

Proof: 1:1½::2:X

1X=3

X=3'' or 80mm

9. To draw a fourth proportion smaller than three given lines:

Reverse the procedure for problem number 8. A is now the longer line and is still marked off first.

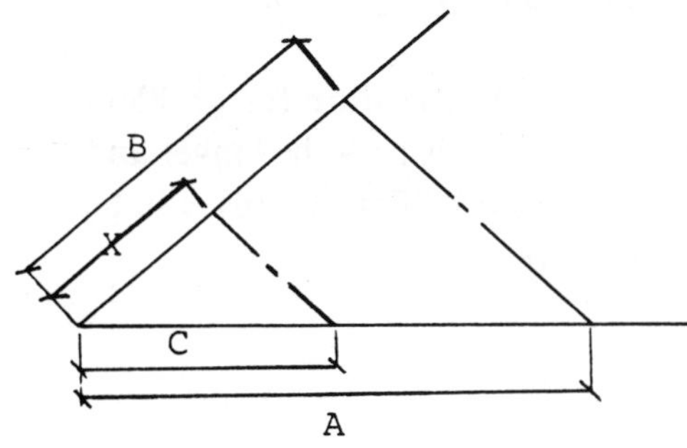

C	1⅛"	30mm
B	1½"	40mm
A	2¼"	55mm

Fourth Proportional Smaller

Proof: 2¼ : 1½ :: 1⅛ : X

2¼x=1.6875

x=.75" or ¾" or 21.818mm

10. To draw the third proportion greater than two given lines:

Let AB be the given lines. Draw two lines at an acute angle to each other. On the base line mark off A and B. B must be used twice, so mark it off again on the other angled line. Join A to B as shown. A parallel to this from the first B will enable the unknown to be measured as 4½".

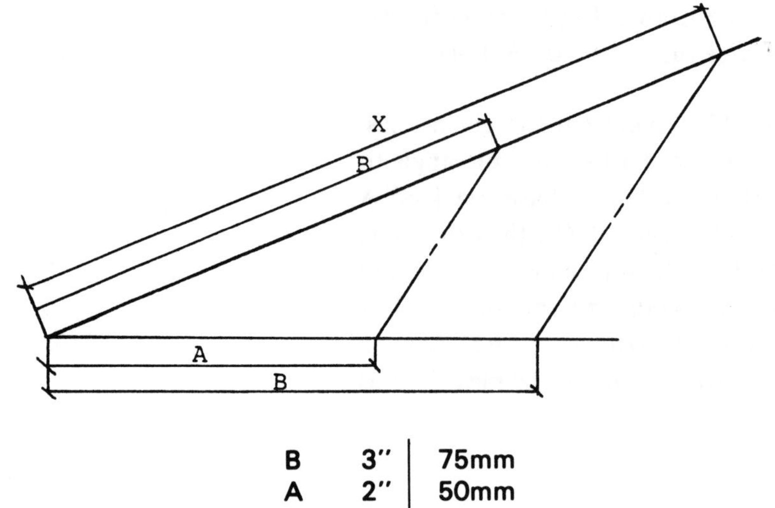

Third Proportion Greater

Proof: 2 : 3 :: 3 : X

2x=9

x=4½" or 112.5mm

11. To draw the third proportion smaller than two given lines:

Reverse the procedure for problem number 10. A is now the longer and is still measured first. B must still be used twice.

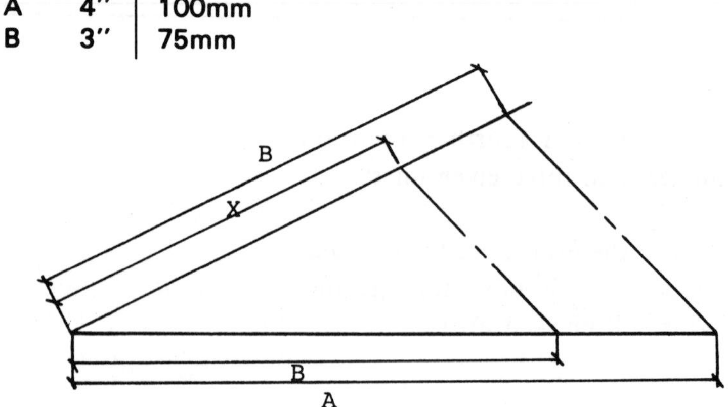

Third Proportional Smaller

4 : 3 :: 3 : X

4x=9

x=2¼" or 56.25mm

12. To draw the mean (between) proportion of two given lines:

Let AB and CD be the given lines. Draw a line the length of AB + CD, find its centre and scribe a semicircle. At BC, erect a perpendicular to cut the semicircle at E. The measured length of x equals the mean proportion.

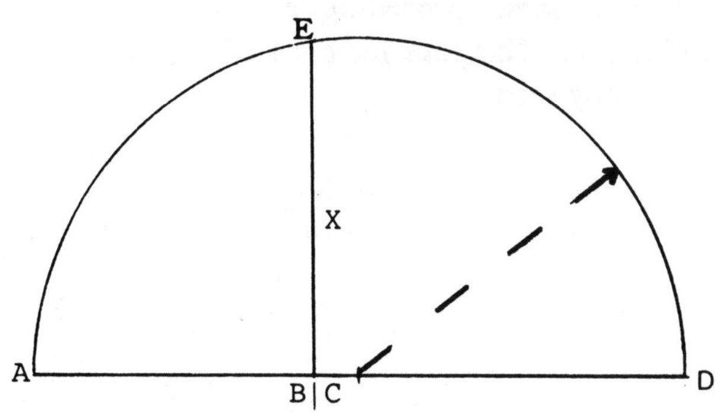

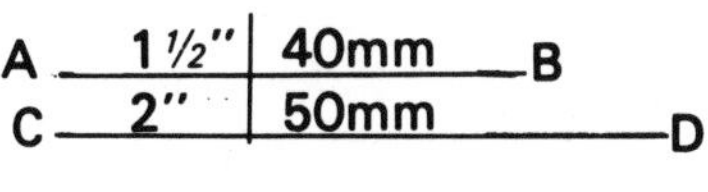

Mean Proportional
AB:X::X:CD
X^2=A.B.C.D.
X=$\sqrt{\text{A.B.C.D.}}$
=1.73" or 44.72mm

13. The square root of a number may be found by the above method if two factors of the number are used:

If a number does not factor easily, 1 can always be used. 5 x 1 are factors of 5. Draw a line 5 + 1 units (6) long. Erect a perpendicular at 5 units and measure it for the root.

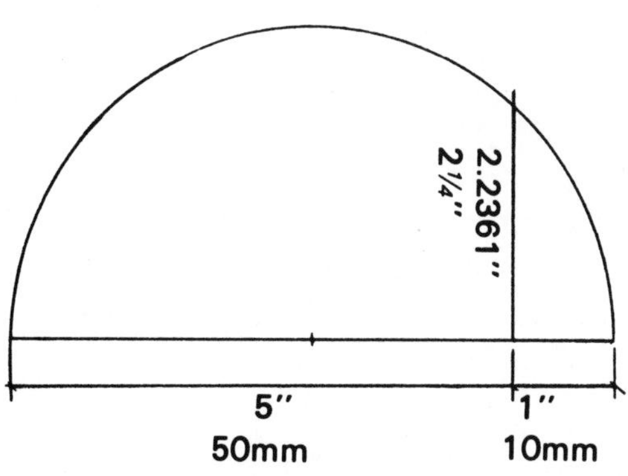

Square root of 5|500mm
use 5 units and 1 unit = 6
$\sqrt{5}$ = 2.2361
by measurement, just under 2¼"
or 22.36mm

14. The same procedure as in problem 13. This time factors 4 x 2 are readily seen.

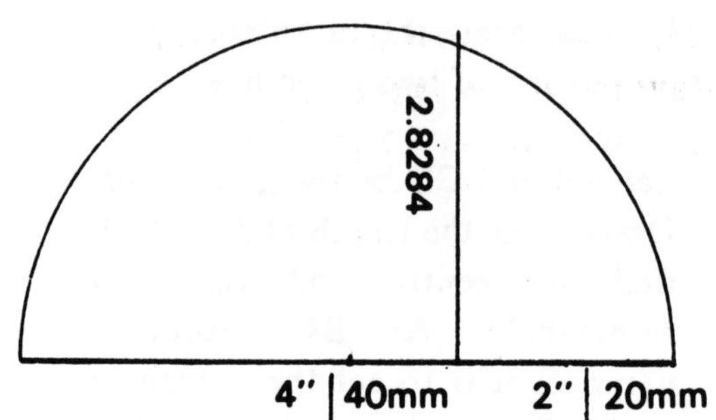

Square root of 8 800mm
use 4 units and 2 units = 6
$\sqrt{8}$ = 2.8284
by measurement, just under 2⅞"
or 28.28mm

15. To construct a triangle on a given base of 2¼" with angles proportioned at 2:3:4 — 9 units.

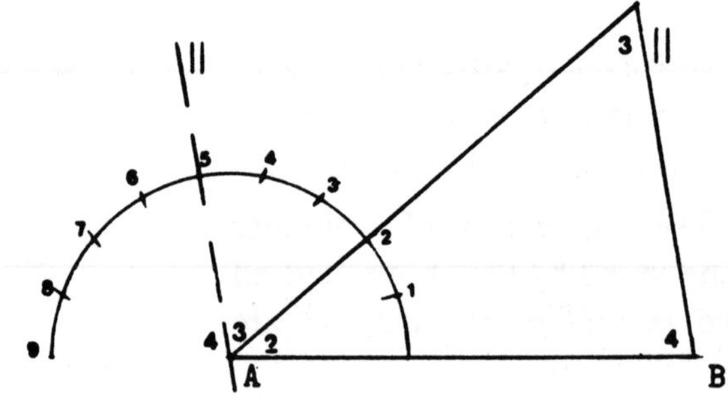

Proof:

$\frac{180°}{9} = 20°$

AB=2¼" | 60mm

2 x 20° = 40°
3 x 20° = 60°
4 x 20° = 80°
180°

16. Given perimeter AB 3¾", layout a triangle with sides proportioned at 4:5:7=16 units.

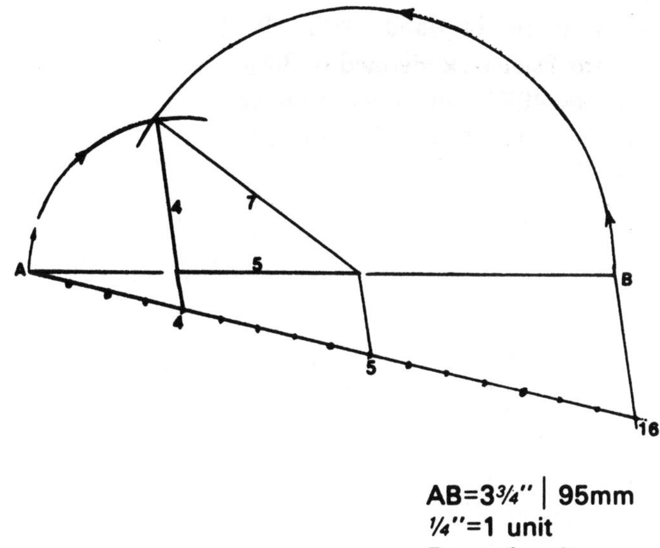

AB=3¾" | 95mm
¼"=1 unit
5mm=1 unit

17. How many pails of stone are needed to mix 9 pails of concrete that has a cement, sand, stone ratio of 1:2:4?

Choose a suitable scale: let ½"=1 unit, 1 unit=1 pail (10mm=1 unit).

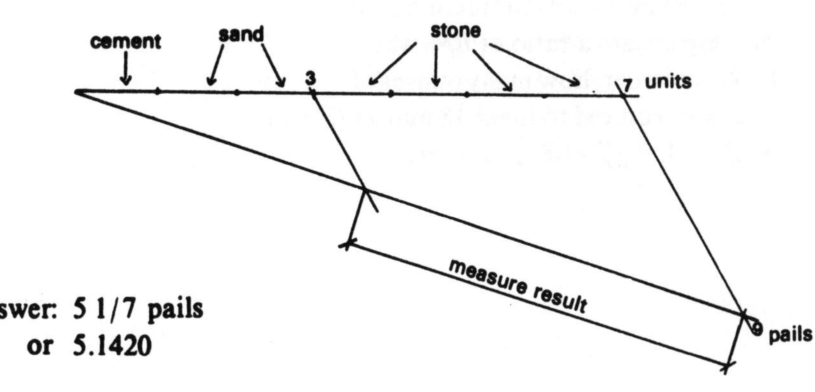

Answer: 5 1/7 pails
or 5.1420

18. It is known to take 10 cubic yards (10m³) of soil to fill a certain 80' (24m) long trench. How far will a 6 cubic yard truck load go? Let ½"=1 cubic yard, ¼"=5'. (read meter for yard)

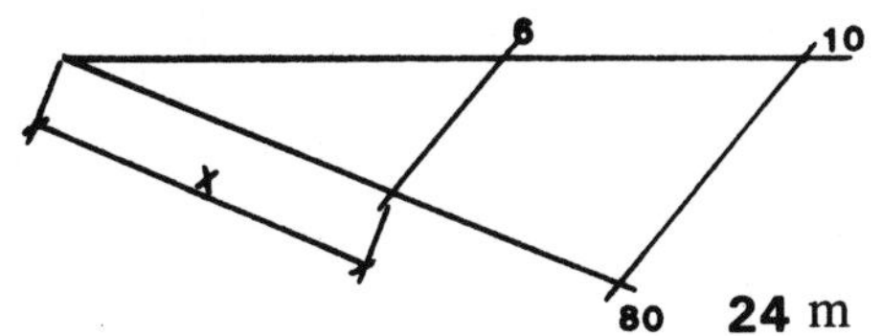

Let 10mm=1m³
5mm=2m

Answer: 48' | 14.4m

Ratio and Proportion

19. A piece of land with 190' (60m) frontage has to be divided to 30%, 30%, and 40%. How much frontage will each lot have? Let ¼"=10', ½"=10%.

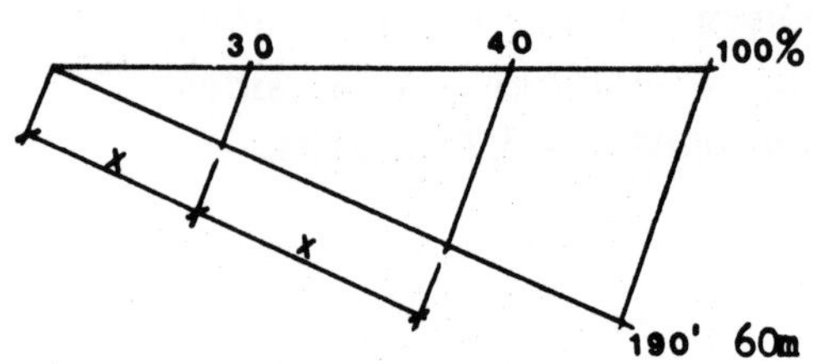

2mm=1m 10mm=10%

Answer: 30%=57' | 18m
40%=76' | 24m

20. If the directions for mixing glue by weight state a ratio of 70% water to 30% powder, how many ounces of each are required to make 18 ounces (500g) of glue? Let ½"=10%, ¼"=1oz.

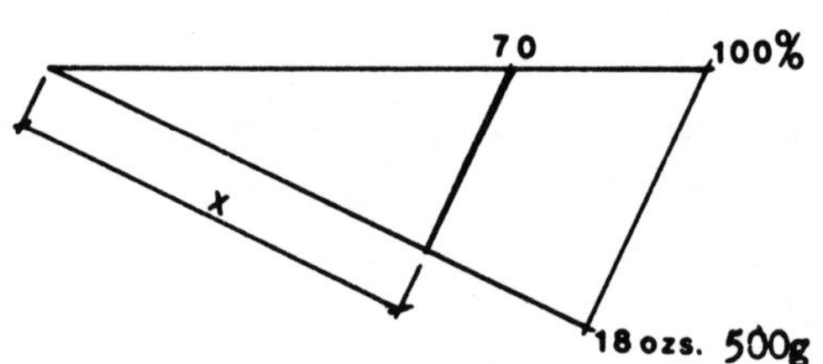

(read gram for ounce)
2mm=1% 1mm=1g.
Answer: 70%=12.6 oz or 350 g
30%= 5.4 oz or 150g

21. A 4' x 8' wall panel that is to have moulding planted on its surface for effect:

Given the long side dimension of 6', what must the short piece of moulding measure to appear proportionally balanced? Draw the diagonal, thus creating a triangle. Measure 6' down the long side and parallel a line with the 4' side. The new length measures 3'. The 3' x 6' dimension can now be spaced equidistant within the 4' x 8' panel.

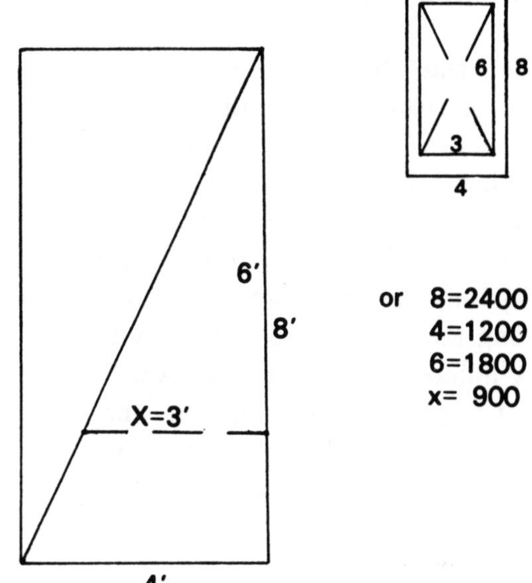

The carpenter's framing square may also be used to solve ratio and proportion problems. The square when layed across a straight edge provides ready made triangles.

22. The sliding of a framing square to achieve the answer in number 21:

Using the twelfths graduations for a scale, set the square on a straight edged board with figures 4" and 8", mark a line along the 8" side. Slide the square down this line to read 6". The new dimension can now be seen on the adjacent side 3", which=3'.

Metric
Or divide metric measure by 10 eg.
2400 = 240mm
1200 = 120mm
Slide to 1800 = 180mm
Answer reads 90x10 = 900

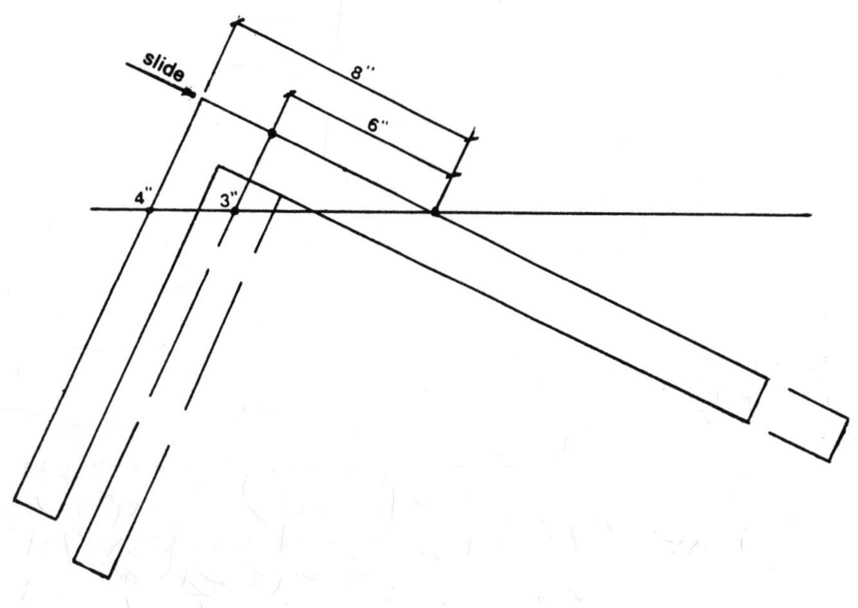

23. Solving a glue problem:

The glue must be mixed by proportionate amounts, 35% powder to 65% water. Therefore, for a total mix of 16 ounces, let ½" represent 1 ounce and 1" represent 10%. Setting the square to 10"=100%, and 8"=16 ounces, slide along the % side to 6½"=65%. On the adjacent edge, read just over 5 3/16=10 3/8 ounces; or, on the tenths scale, 10.4 ounces. The remaining 5.6 ounces of the total=35% powder.

let:
16oz=450g
2mm=1%
1mm=1g
65%=292.5
∴ 35%=157.5 powder

Metric
Set up 225mm (grams) and 200mm (%)
Slide to 130mm (65%)
Answer reads just over 146x2 = 292.5

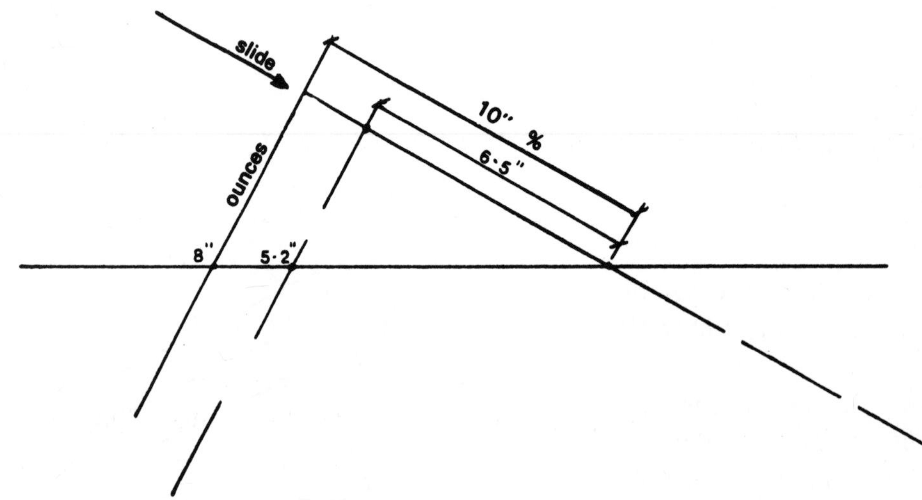

The above are fourth proportion problems. Many problems may be solved this way, to a fair degree of accuracy, by letting a scale represent the problem quantities.

Discovery Exercises

Use drawing instruments to solve the following problems. State all measured answers.

1. Determine the fourth proportion greater than the given dimensions.

(a) 2"	**50mm**
(b) 2½"	**64mm**
(c) 2¾"	**70mm**
(d) X	

2. Determine the fourth proportion smaller than the given dimensions.

(a) 4"	**100mm**
(b) 3"	**76mm**
(c) 2"	**50mm**
(d) X	

3. Determine the third proportion smaller than the given units.

(a) 8 units
(b) 6 units
(c) X

4. Construct a similar triangle with sides proportioned at 2/3 of the one shown. State new dimensions ± 1/32" (1mm).

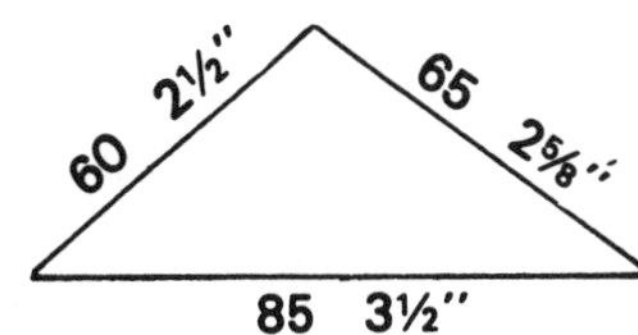

5. It is permissable for a certain job to use 5% clay and 15% stone fill, but the bulk must be sand fill. Show by geometry what proportion of sand would be contained in 24 yards^3 (24m^3) of fill. State the answer to the nearest $^1/_4$yard^3 (.25m^3).

6. A small building of 1,800feet^2 (195m^2) has been designed as shown with areas A,B,C,D proportioned at 2-2-4-5 respectively. Show by instrument drawing the footage (m^2) alloted to each area.

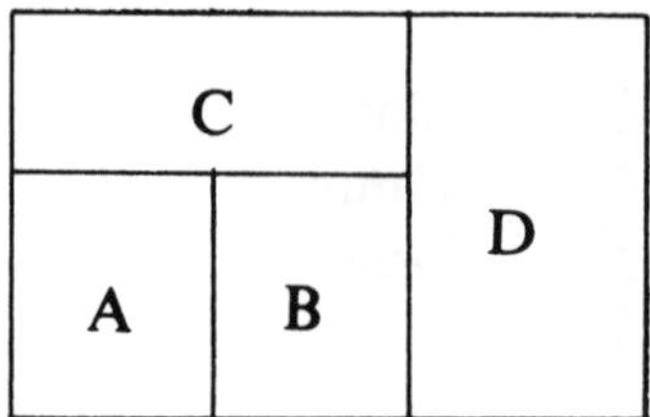

7. 6yards^3 (6m^3) of concrete are needed to pour a basement floor. If a mix proportion of 1-2-4 (cement, sand, stone) is used, estimate the number of bags of cement required for costing. Use one 80 pound (40kg) bag as 1 cubic foot (.03m^3).

8. Determine the mean proportion of two items weighing 200 pounds (kg) and 350 pounds (kg).

Unit 8 Stairs

GEOMETRIC APPROACH TO STAIRS

It is assumed that a student has an understanding of the basic principles of stair construction before starting this unit.

1. In the diagram, a total rise and total run are divided into 15 unit steps for a straight flight of stairs. By studying this drawing in terms of similar triangles, it can be seen that the well hole or headroom, if unknown, would present a ratio and proportion problem. With known information, either the headroom or well hole dimension could be found by sliding a steel square.

Example: to find well hole.

a) Set a square over a straight edge with the unit rise on the tongue and the unit run on the blade.

b) Slide the tongue to the scaled dimension of desired headroom plus floor thickness.

c) Read the scaled well hole on the blade. Reverse the procedure to find the headroom, remembering to allow for floor thickness.

These problems and others can also be solved by layout on a pitch board. The drawing shows the application of a pitch board as compared to a steel square.

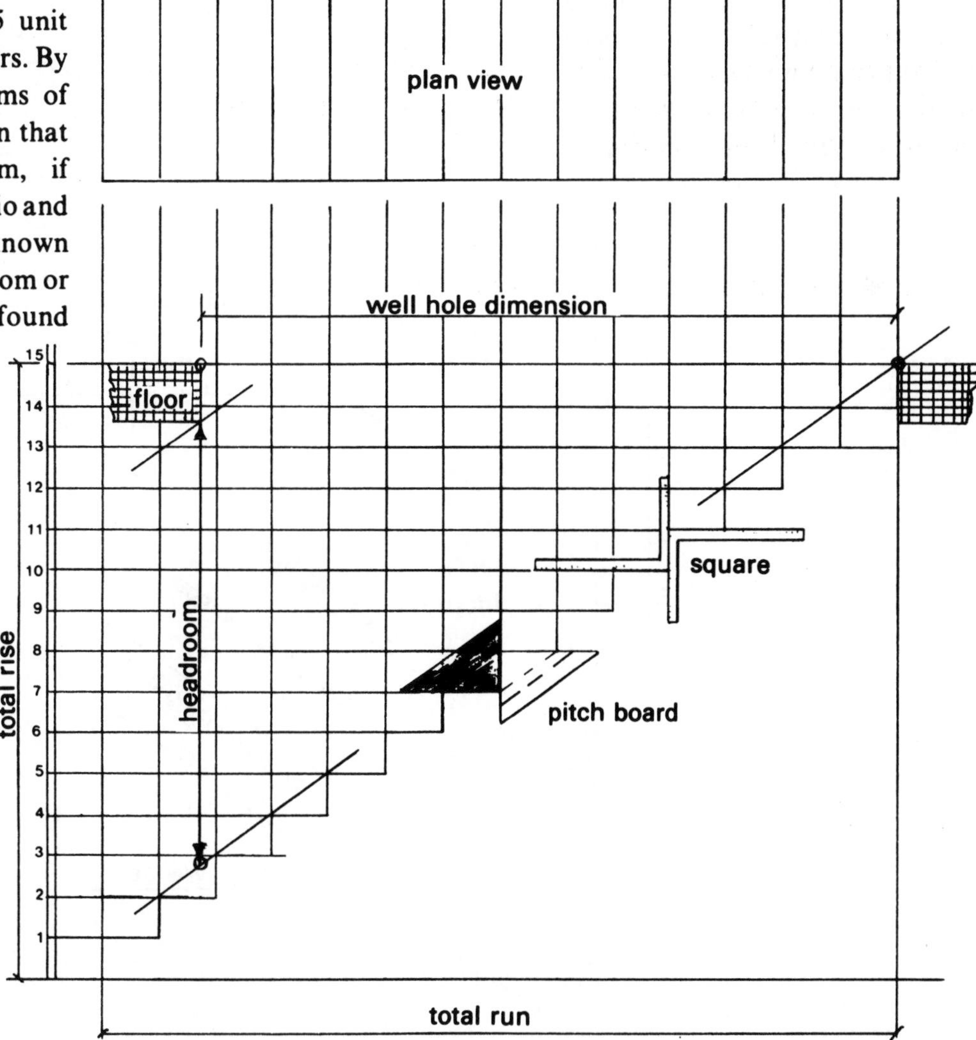

2. Pitch board:

The pitch board is made from plywood or masonite and is a permanent record of the stair job, should it need to be reproduced. Study the layout carefully; the template must be a perfect right angle. The pitch of the stair is shown at p. A diagonal is drawn through a rectangle made by the tread and riser thicknesses shown at x. Extend x and measure ½" (12) for wedge allowance. (z and y represent the scaled well hole and headroom, drawn parallel to the pitch line.) Drawing 'a' shows the pitch board being used to layout a housed string from the bottom edge of string. In drawing 'b', the guide has been renailed at the cut string line and the layout is done from the top edge of string.

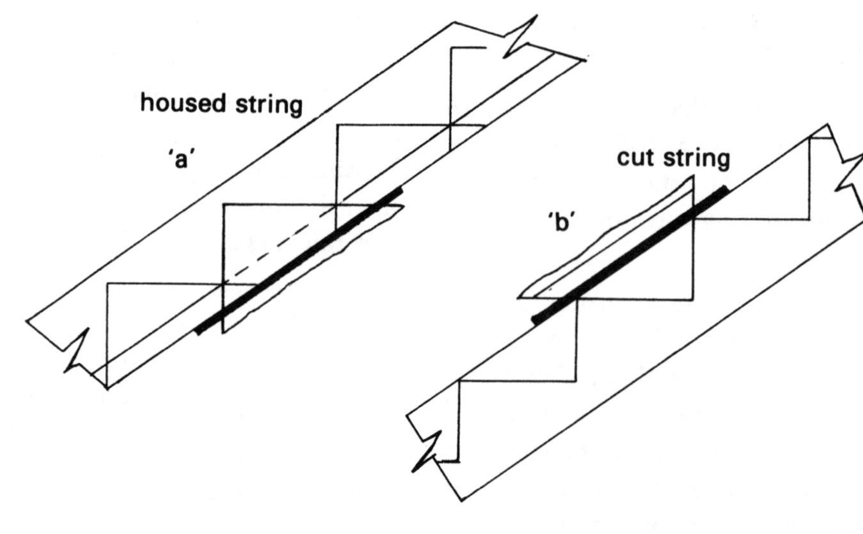

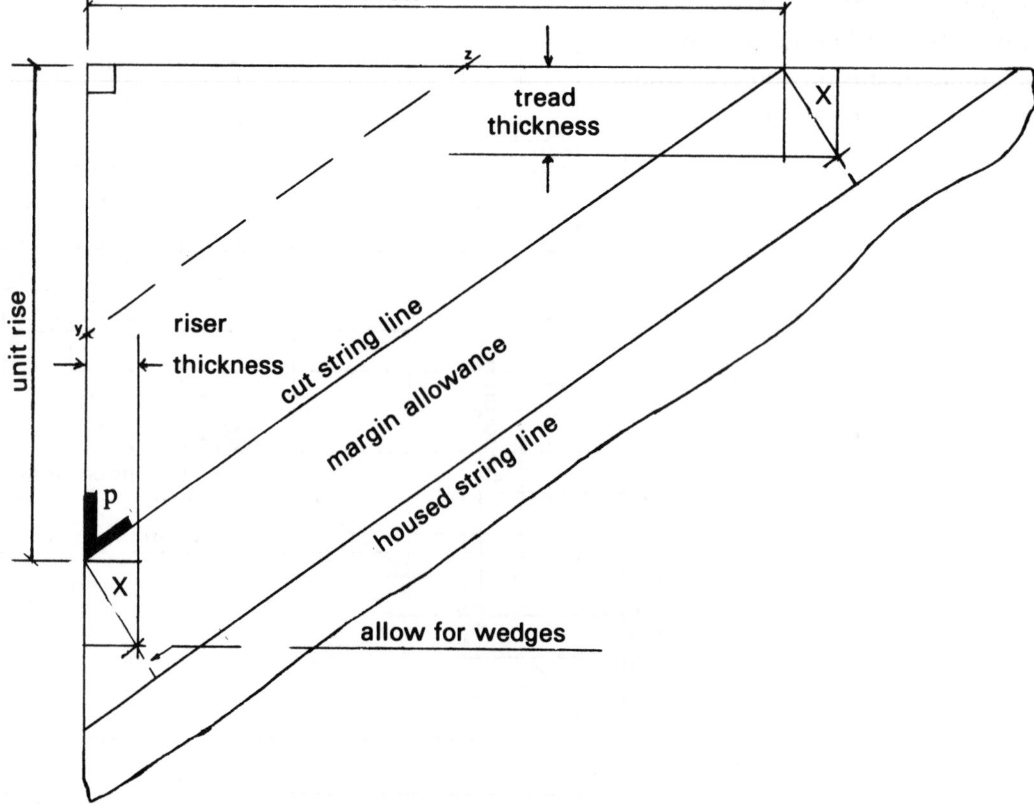

3. In this diagram, the pitch board is used to determine the difference in height of ballusters. With the ballusters being layed out across the unit run, A is the difference in height (length). b is a plumb cut and is shown transferring the handrail height around a newel post and giving the cut for balluster to handrail.

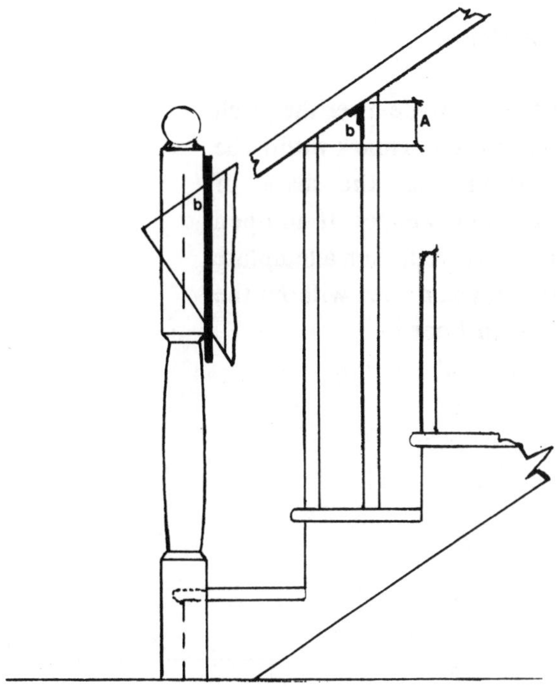

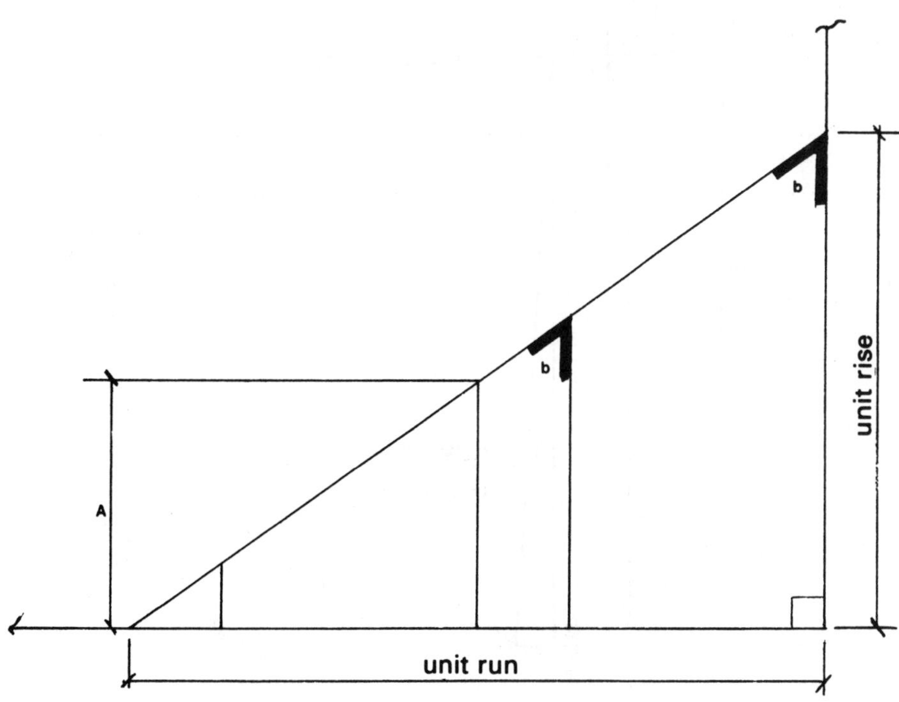

4. Template:

This pattern is made after the pitch board and is used to mark within the housed string lines the shape of tread, riser, and wedges. If no open strings are involved, then a template could be made and used without the aid of a pitch board.

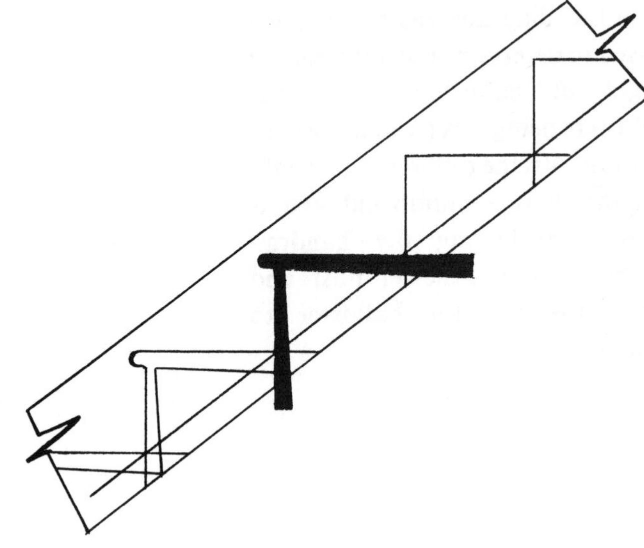

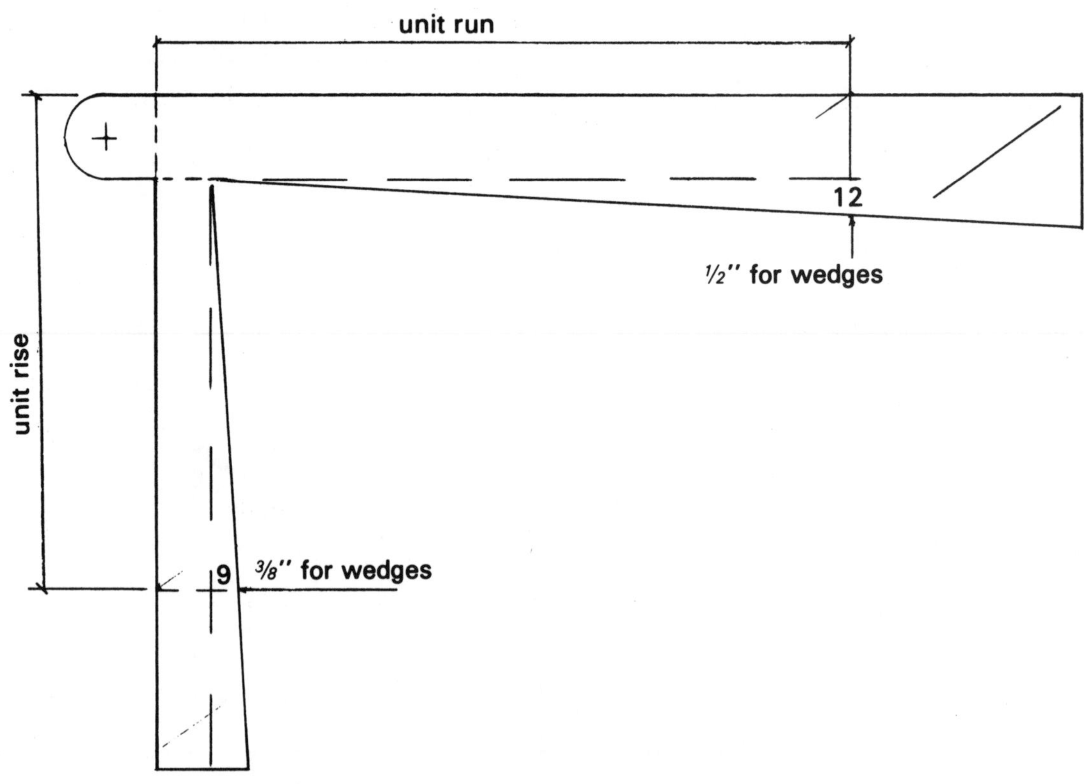

(study p. 73-78, then do this layout)

5. Layout of a stair with an open string, a housed string, a bullnose starting step and newel posts:

Start with the housed string. Mark in the required margin line. Using steel square or pitch board with calculated units, lay out a floor line and plumb line. Consider the possibility of a baseboard butting to the plumb cut, hence leave enough string material. Starting at A, layout the desired number of unit steps. Once again consider a baseboard junction at the landing. An option is to cut the string back level below the landing nosing.

Notice the cut string starts at the second riser, which in this case does not make contact with the floor. Lay out a plumb line B, then the number of units between the newel posts. Pay particular attention to the termination at the top. The tenons top and bottom are measured in from riser face, which are therefore deducted from the unit run. (Usually tenons pentrate half of newel.) The riser face aligns with centre of newels.

Use 8 risers
unit rise 7" 180
unit run 10" 250
scale 1"=i'-0" |1:10

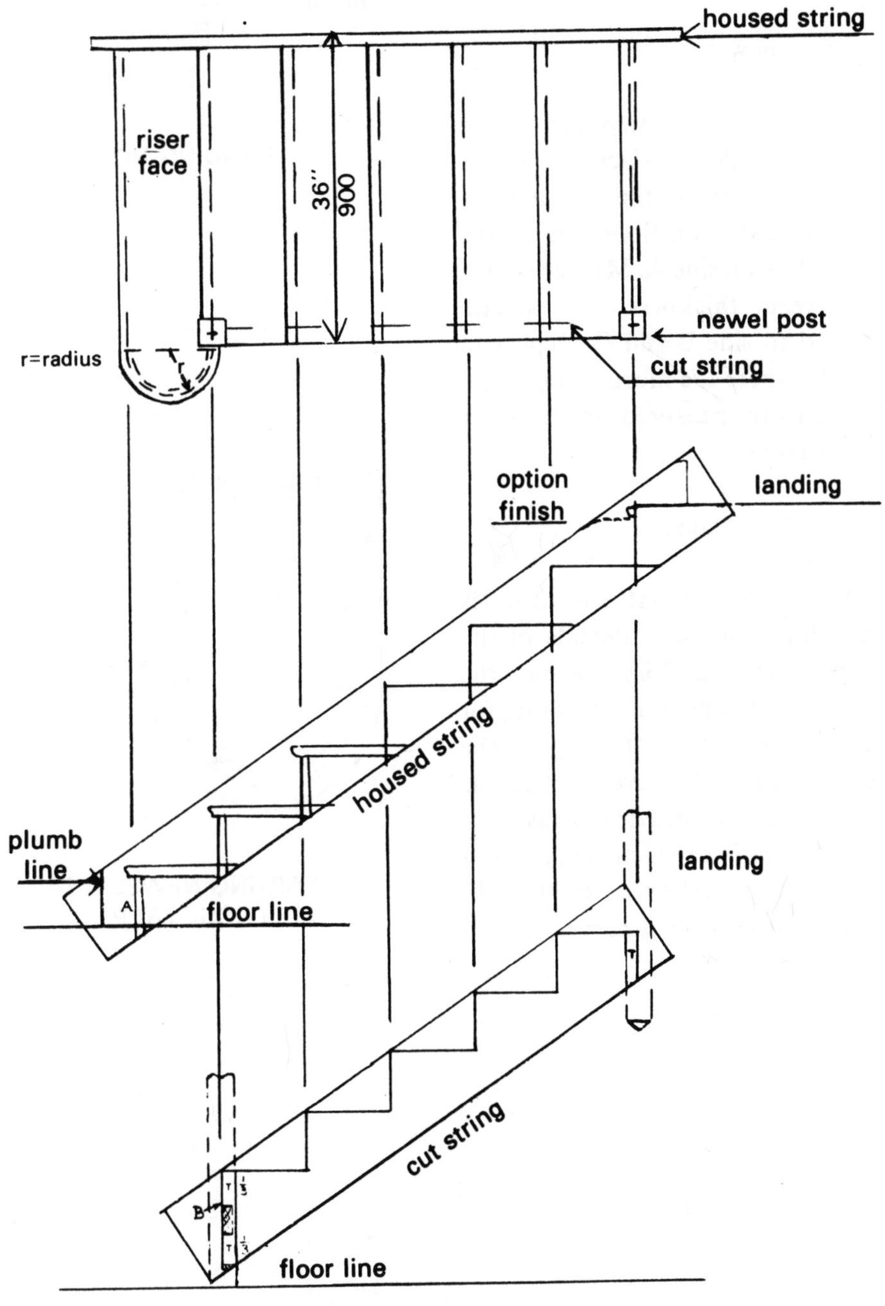

6. Landing newel:

Leave a desired amount (usually 3"-4") of newel extension at the bottom, then start the layout on side B. Layout for tenon as per string, then the tread around to side C. Measure up the unit rise and, lastly, the landing nosing which shows the depth of dadoe ½" on side D.

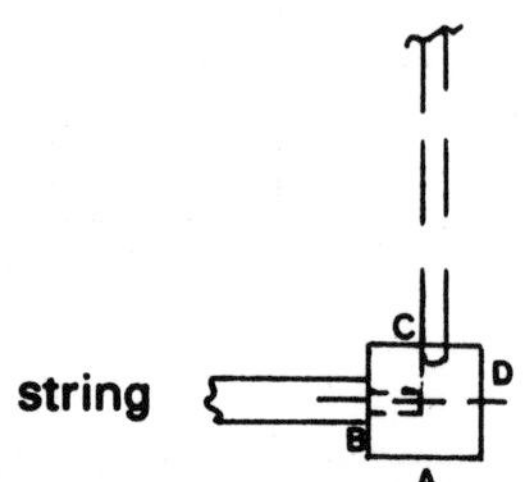

LANDING NEWEL

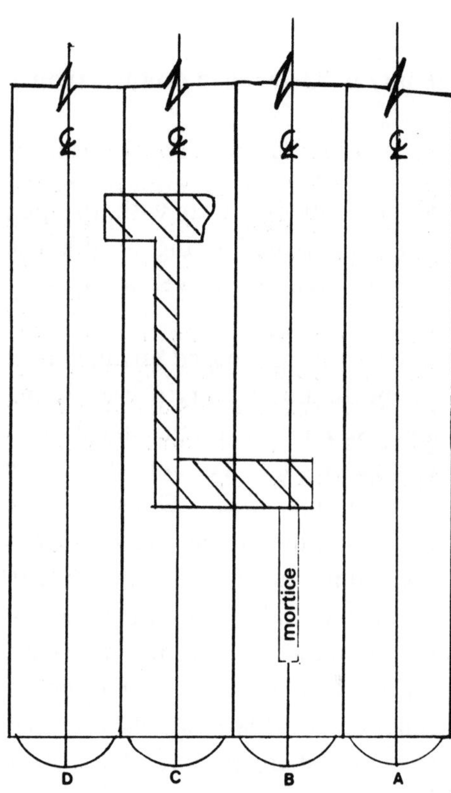

Starting newel:

For the bullnose step the layout starts on side A. Face of riser is center of newel, so the first unit rise is measured from floor line up the centre line on side A. Measure down the tread thickness and square around to side B and C. Repeat to side D where the mortice holes and haunch are measured down as per string layout.

The bullnose step:

This layout is straightforward and is best done on the underside of the tread material. First layout the newel post position allowing for dadoes. Half the unit run gives the radius point of curve. Notice the center point is plotted ½" away from newel face. The reason being, that it is easier to join the riser at 90 to the newel, rather than a curve.

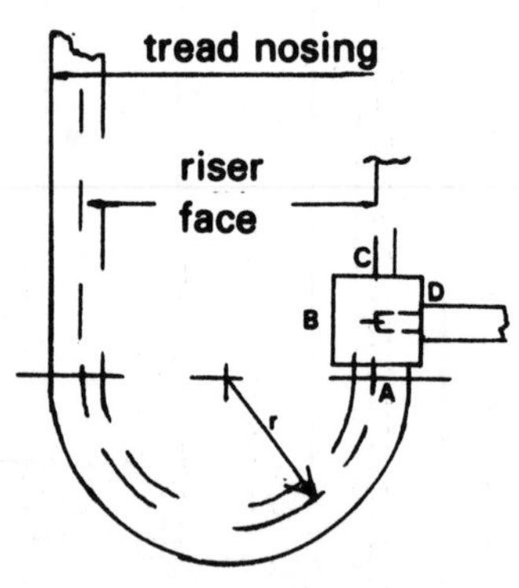

STARTING NEWEL AND BULLNOSE STEP

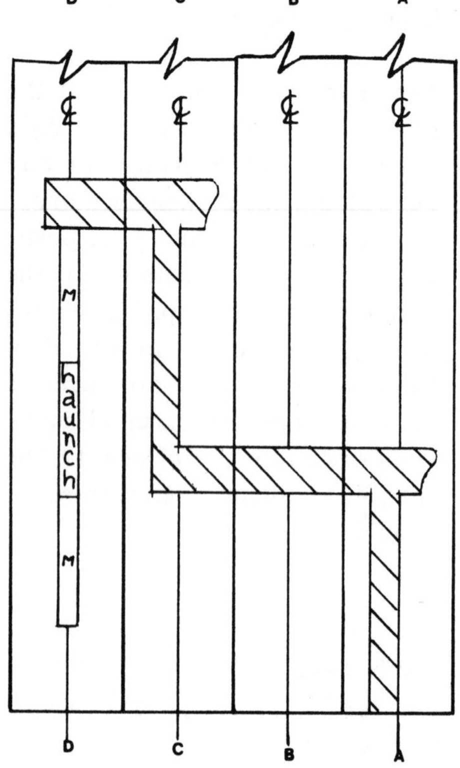

7. Quadrant step:

First layout the newel post position. Actually only a little more than a quarter of newel will show. The spring line for the curve is drawn ½" away from the newel to facilitate a better riser joint. The radius is the distance from the spring line to face of riser. Measure this distance from M to N for center point. The newel layout is as shown.

QUADRANT STEP AND STARTING NEWEL

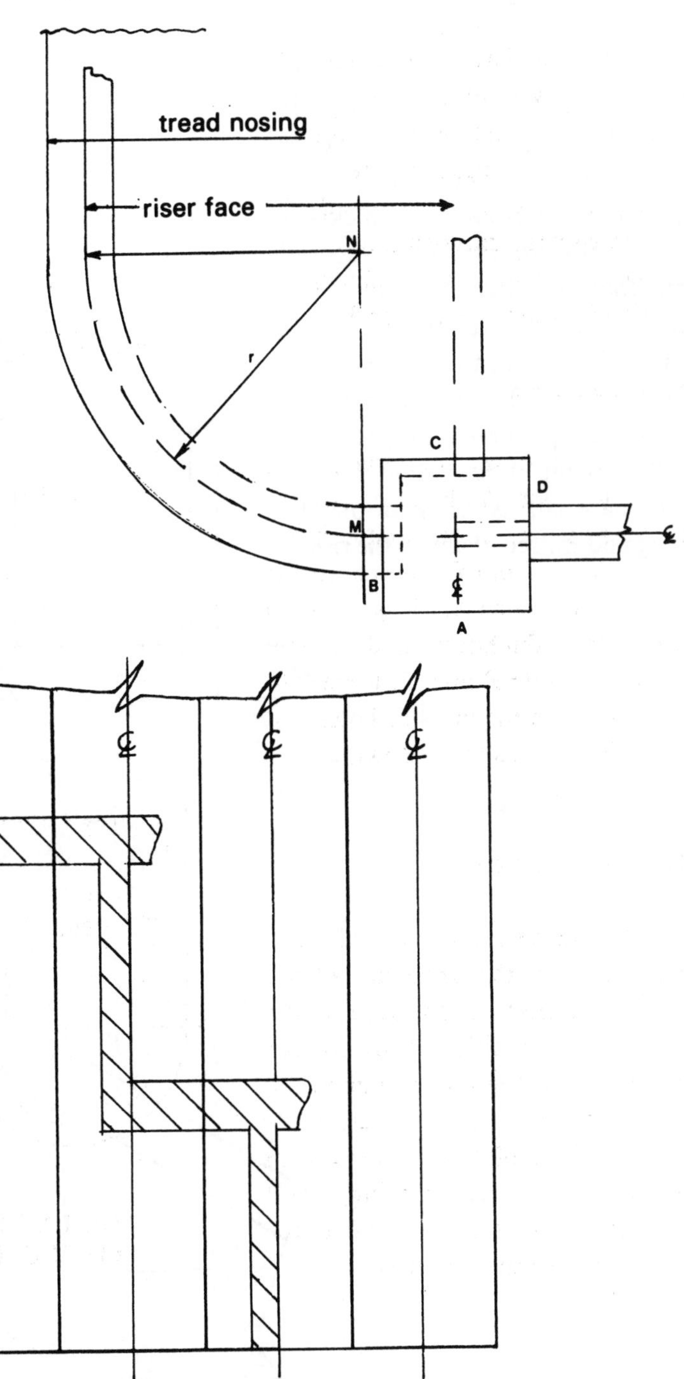

8. Riser to open string:

The optional joints are shown at E, F, G. E is a 45° joint which is acceptable, but does not have as much surface contact as F. To layout for F, measure riser thickness A across the unit run, plumb up and over the top edge of string. The diagonal is the mitre cut. After cutting the string, this angle can be transferred by T bevel to risers. G shows a simple bracket layout. The bracket, being about half as thick as riser material, is planted on the string and mitred at 45° to the risers.

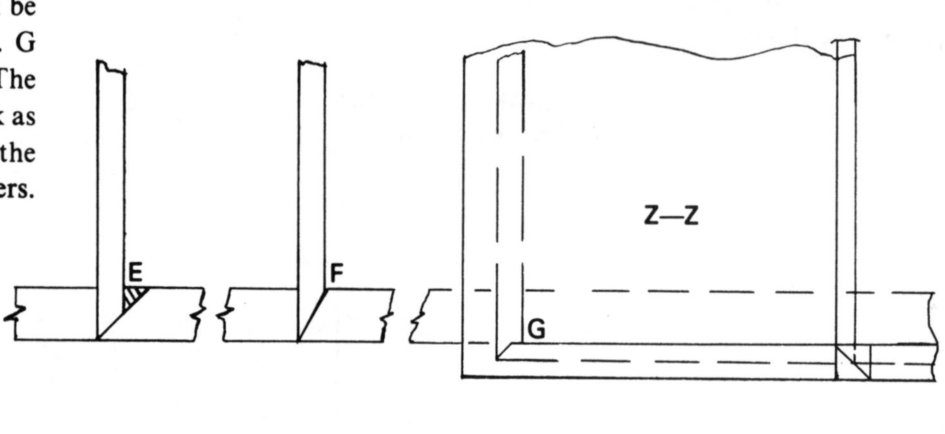

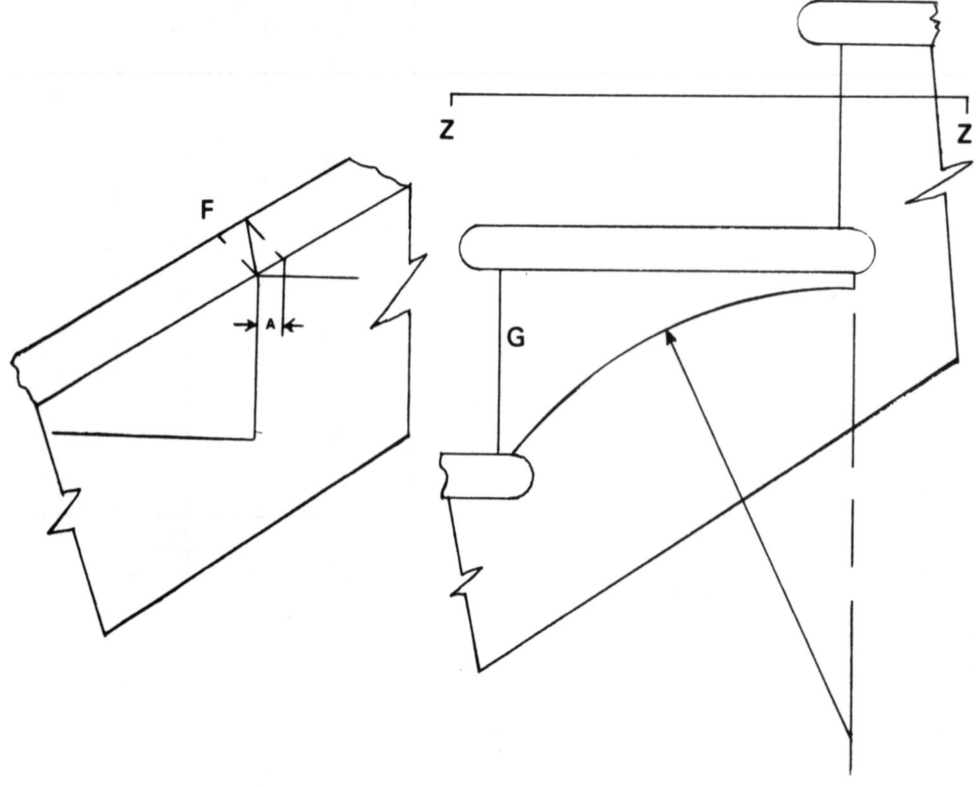

9. Winders:

This type of stair is used to gain height within a limited area. When used, the turn should be 90° and at the bottom. A plan layout of the wind is necessary. Draw a walking line 15" from the newel post, the idea of this layout is to keep the three winders as near as possible to the unit run at the walking line. Often a centre will have to be established outside the newel to achieve extra tread width. Next, the outer strings must be developed. Lay off the first unit rise. Project the tread width ab to a^1b^1, the second unit rise, then project c to c^1. The middle tread (kite winder) also shows on the upper string. Now the string can be drawn in about the steps showing what shape and width of material is required, some approximation is acceptable here. The guide lines are:

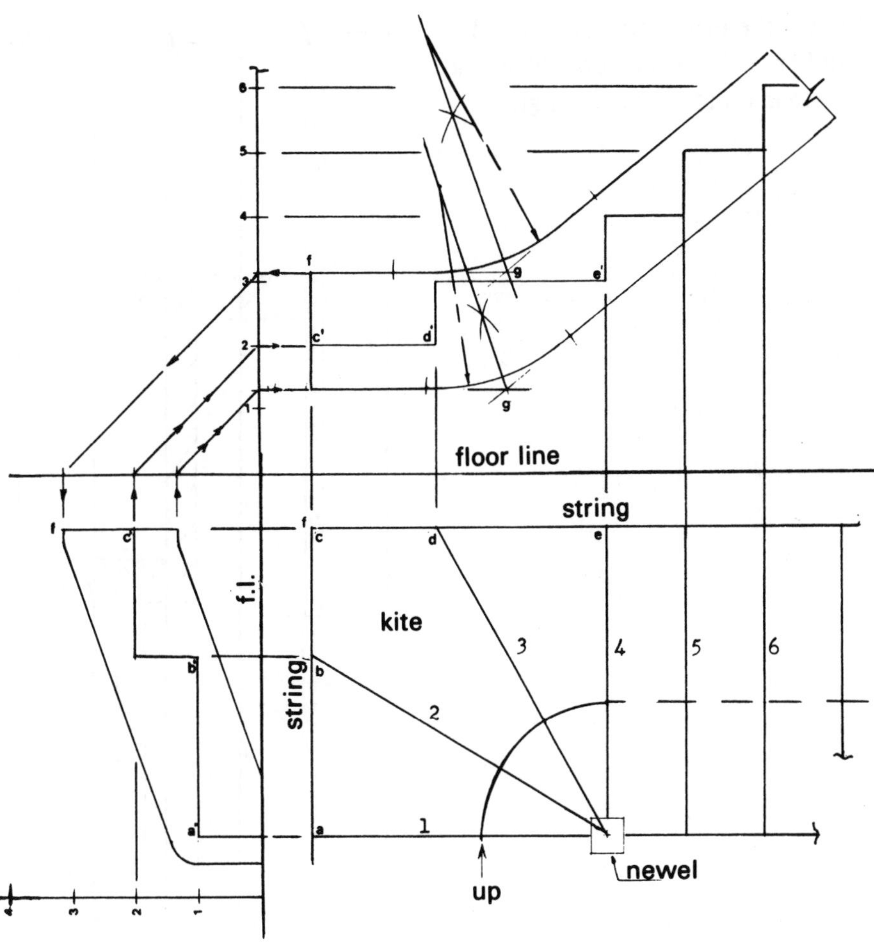

a. Allow sufficient margin and nosing room.

b. Be sure to have equal material above the kite so the two strings meet at f.

c. The top and bottom edges could intersect at g with straight lines, but an easing is more graceful and if desired a suitable curve may be found on the bisection line of g.

NOTE: Find the internal curve first, see direction of arrows.

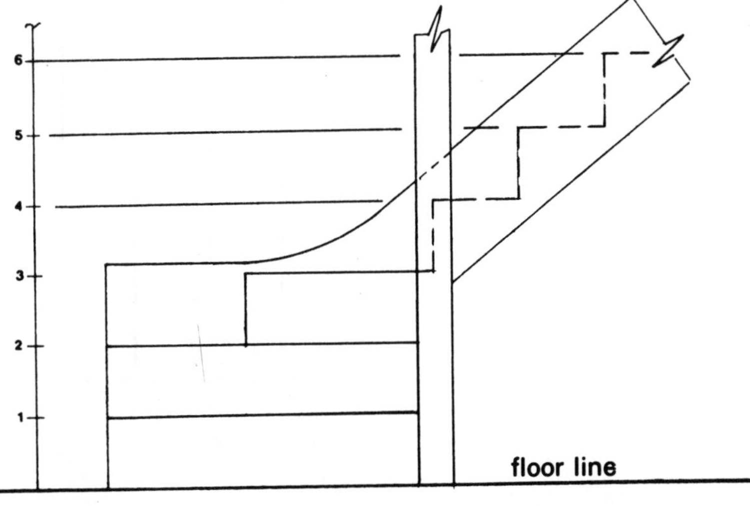

10. Winder newel:

Follow the procedure as outlined in number 6. Keep in mind here that the steps dadoe is at an angle.

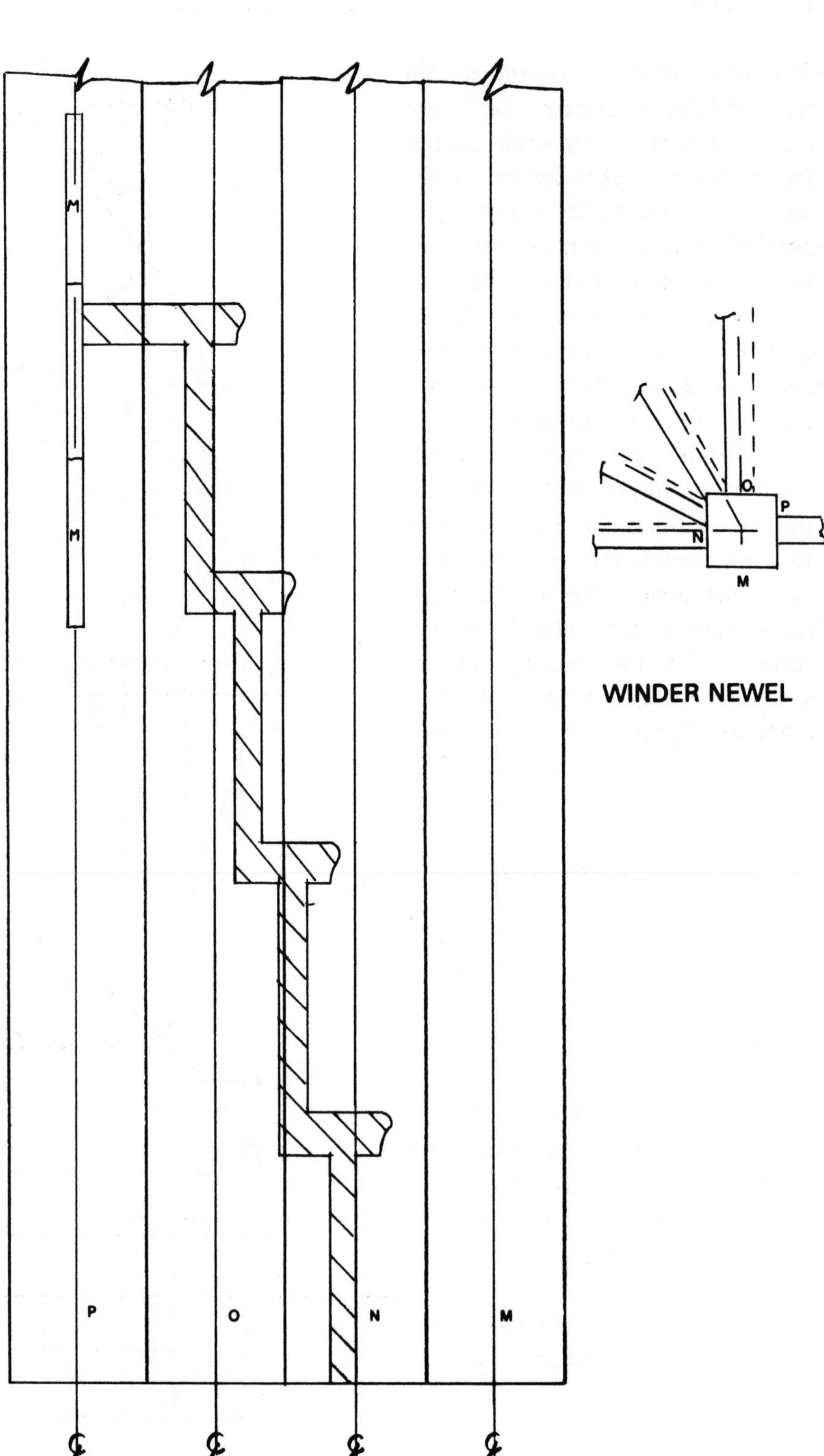

WINDER NEWEL

11. Circular stairs:

All curved stairs follow the same principle in layout and construction. A tight circle such as a spiral or an elliptical plan, needs a little more care and may take longer to complete. This drawing shows the plan of a circular stair and a front view. First, calculate a unit rise then layout the curve. For the curve you need a radius, which may be given, or a well hole size may be given as X and Y. Strike a walking line and divide it equally into the desired number of unit runs. Adjustments may be made to maintain a reasonable tread width. Each riser face is now shown radiating from centre 0. The unit measure to be used for stringer layout is now seen at AB and CD.

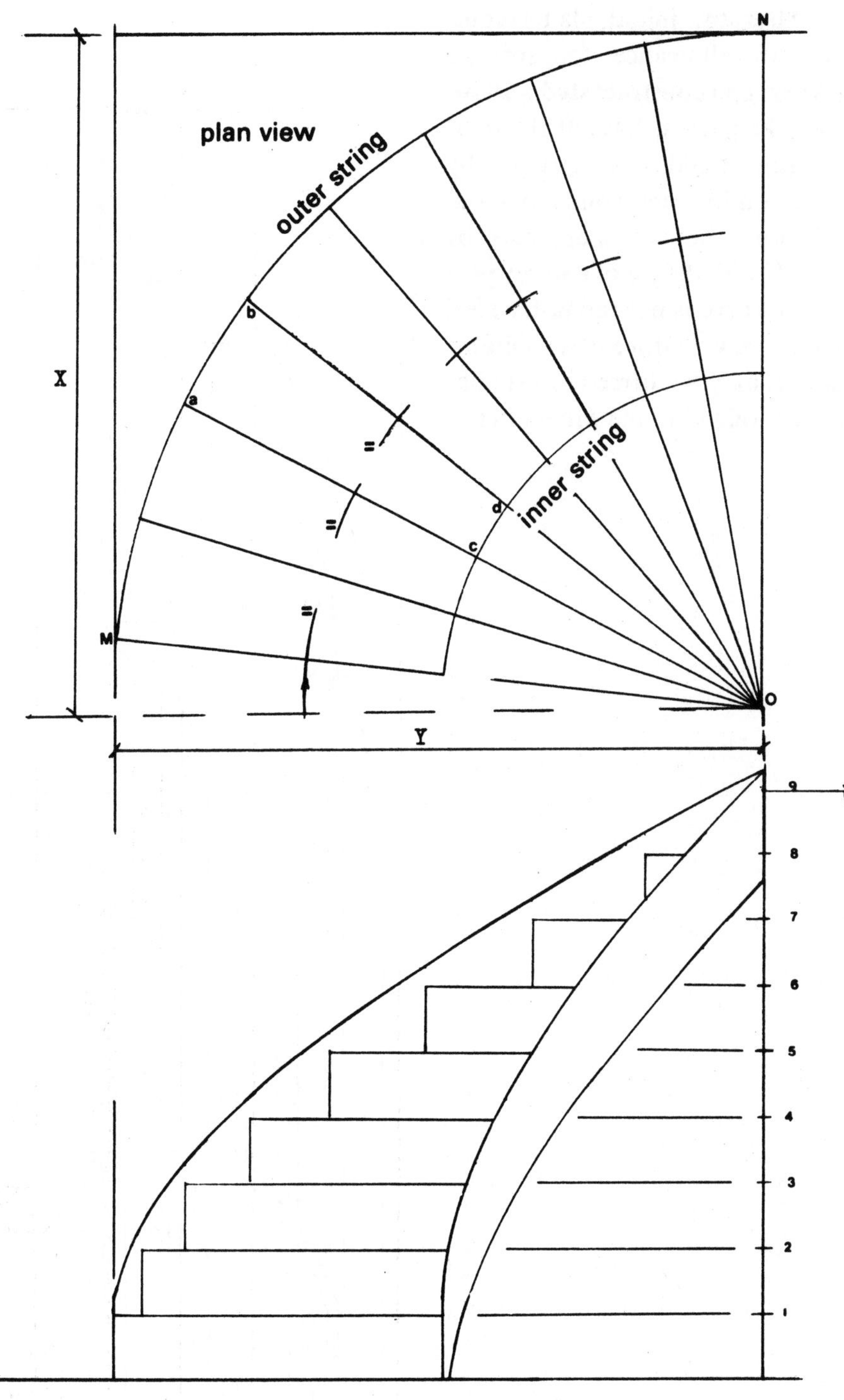

12. After the initial plan layout, make an allowance for stringer thickness and construct stud walls to match the curves. Lay off the total rise and build the strings by lamination between points MN and total rise. The unit steps may be layed off with the aid of a storey pole or if the curve is not too fast, a steel square may suffice. A suitable handrail may be formed in a similar manner following the same pitch.

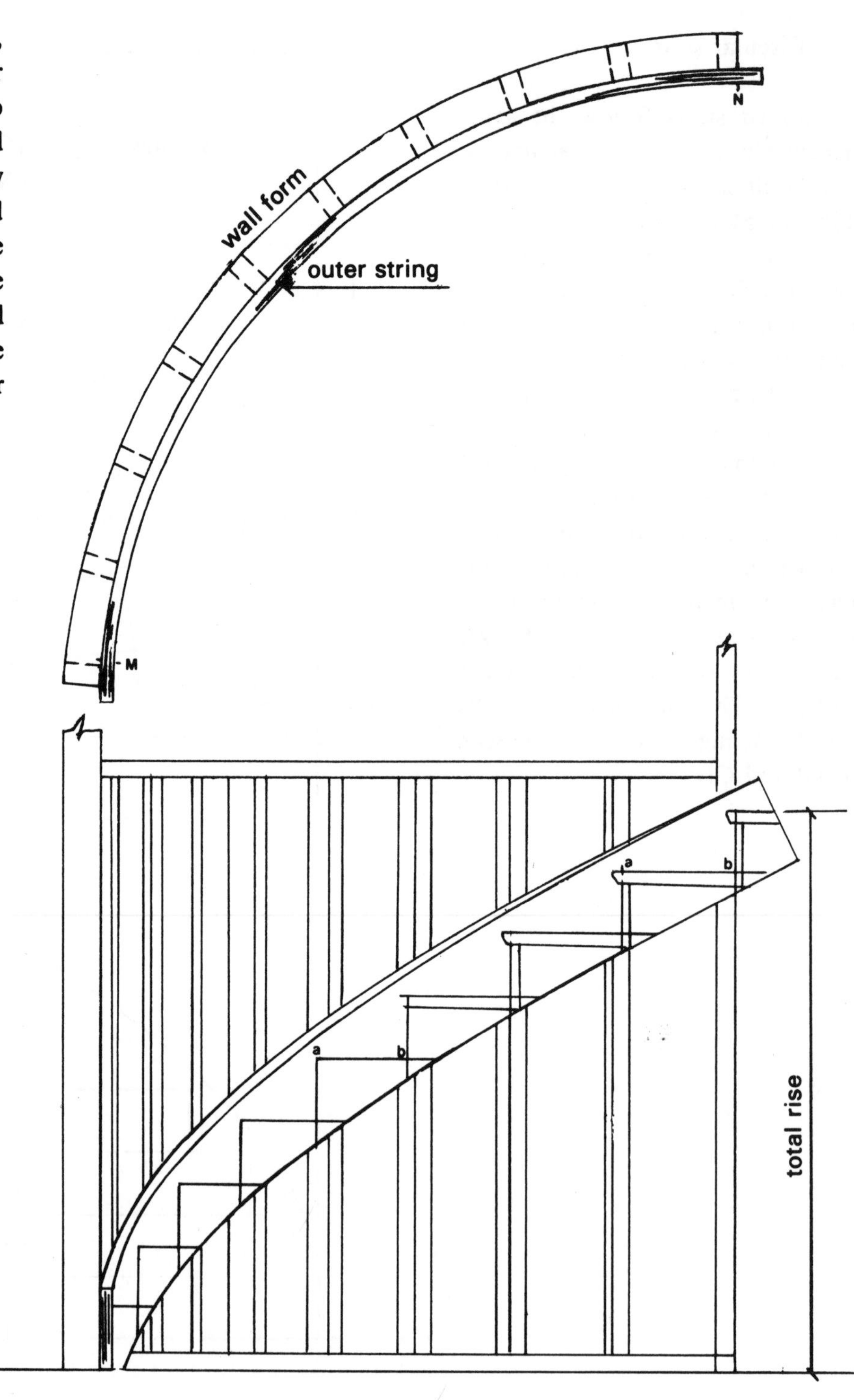

13. Handrailing:

Shown is a simple scroll over a newel post and at the top a goose neck to a level handrail. The radius at the top is found by bisecting the inclination of the stair handrail and the plumb line over the riser. The center is where this bisector intersects a level line extended from the landing rail at 0. Use the same radius on a similar bisector at the bottom.

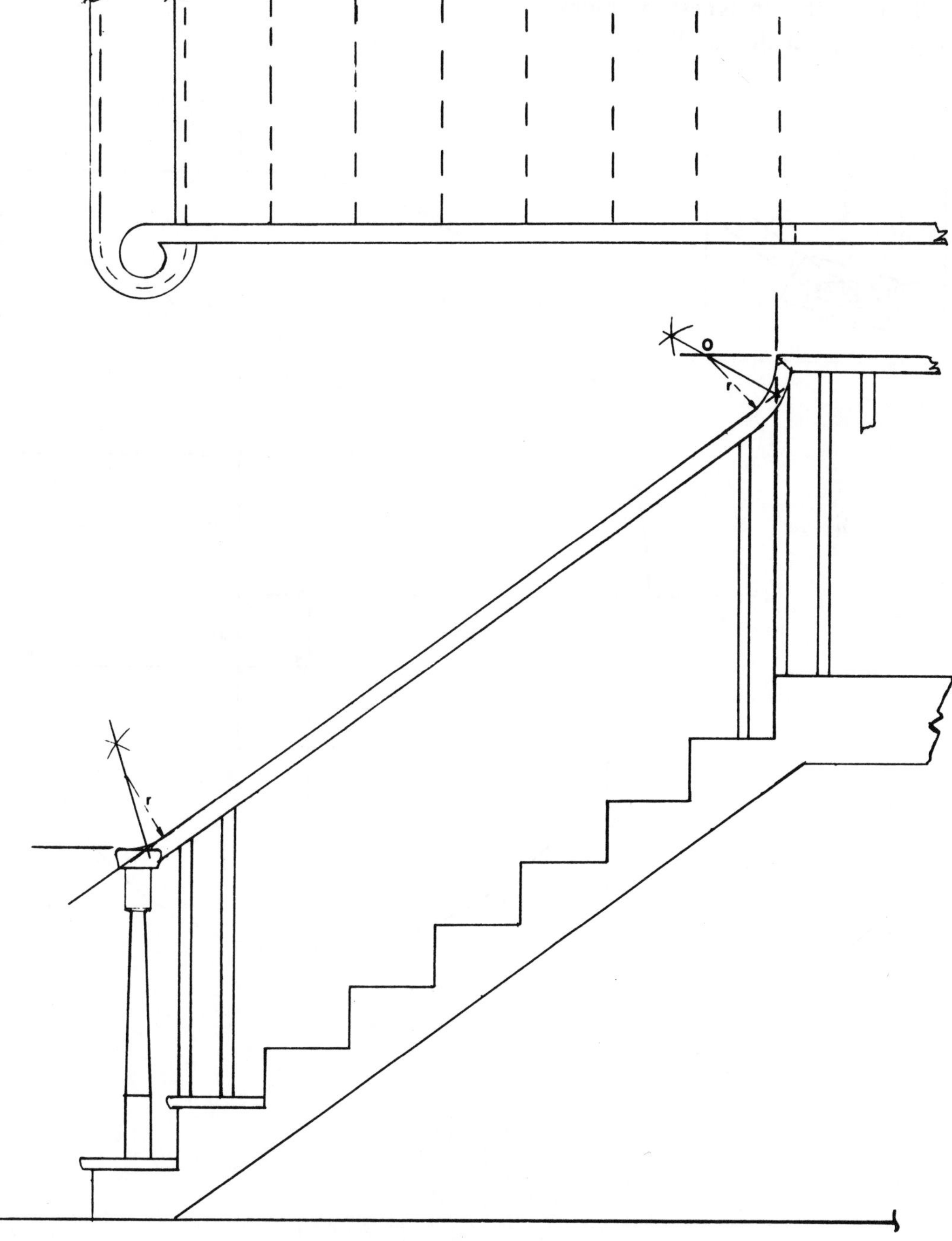

14. Scroll:

To layout this simple volute draw the width of rail W. Next, draw line ab and lay off ½W for the first centre which gives arc cd. Now form a square to give centres 2, 3, and 4. Lastly, decide on a location for the joint j. A frontal layout as per inset will show the thickness of block required to cut the scroll.

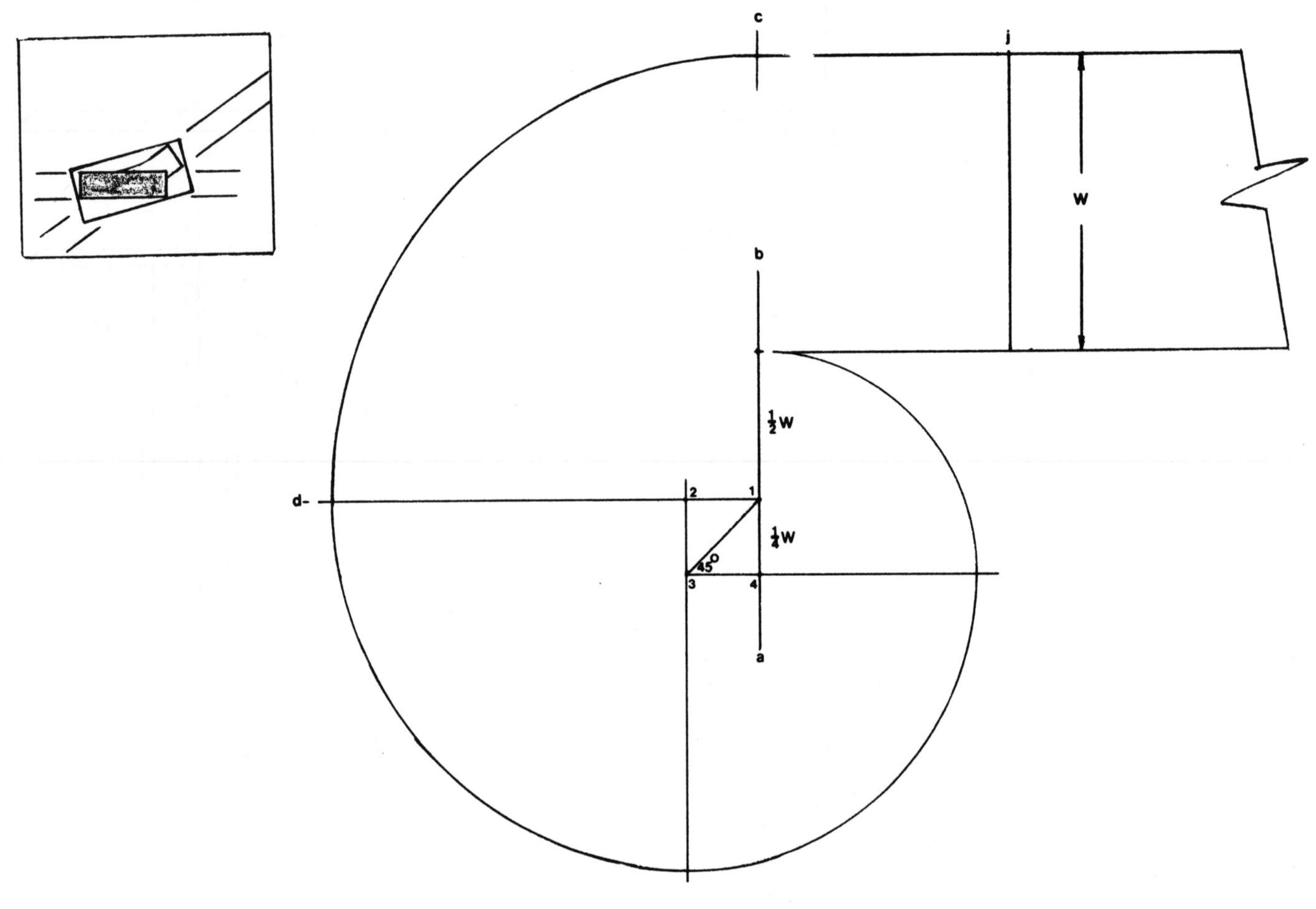

15. A wider scroll pattern may be achieved by this volute using more centres:

Establish a first line ab across the rail. Measure 2w to fix the first centre for arcs cd and ef. Fix centre 2 at ½w from 1. A 30° line from 1 intersects line 2g to give 3, a 45° line from 2 intersects line 3h to give 4. Project from 4 to get 5, and from 5 to get 6.

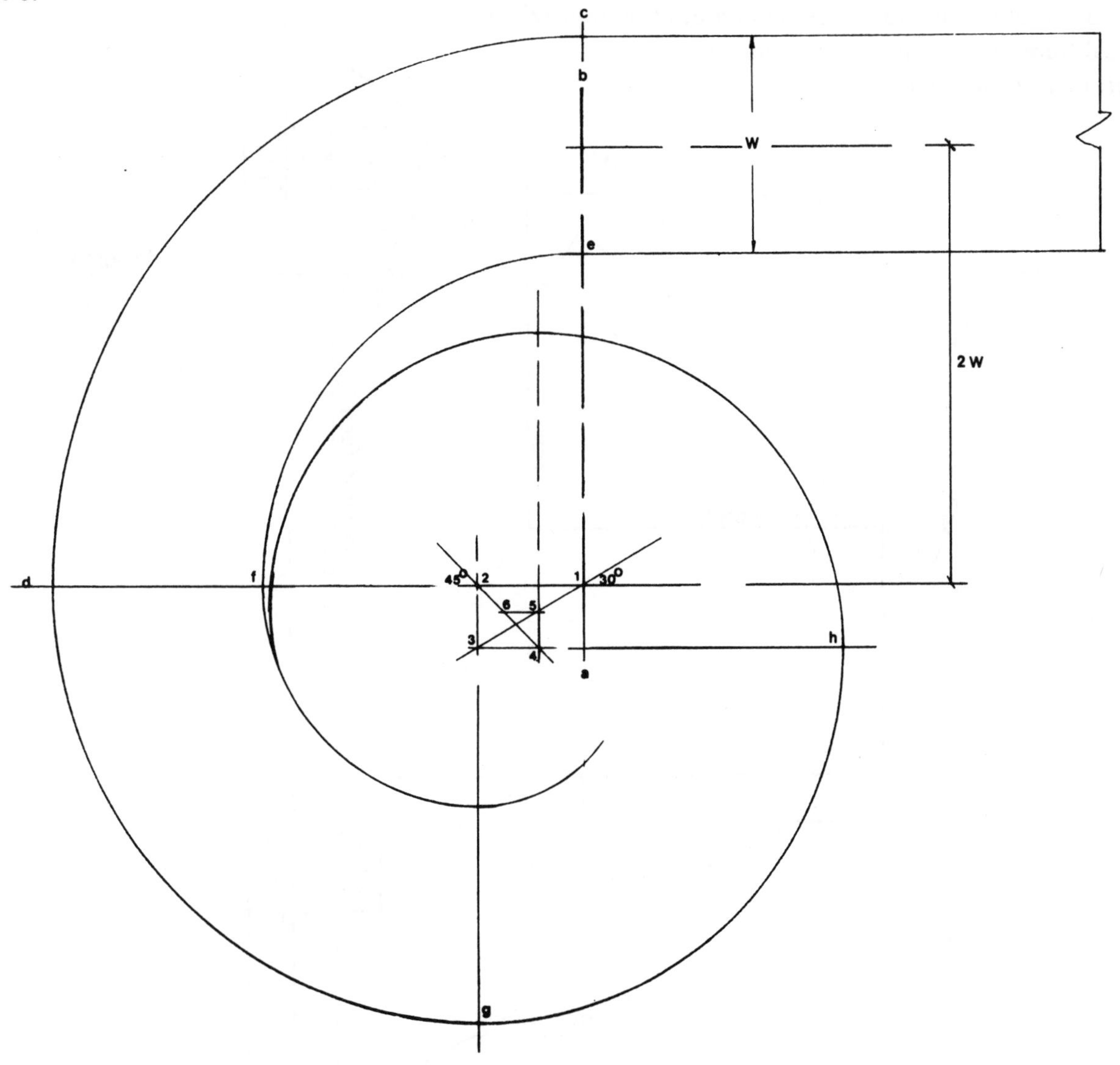

16. The top of a handrail can be varied to suit almost any situation: This diagram shows an "over easement" to meet a "level quarter turn":

Handrails over landings are normally higher than the vertical distance over a nosing line, therefore, a direct relationship exists between the length of easement and quarter turn and the horizontal (plan view) location of landing rail. "X" equals the differential in height. BC — CD represent center lines to the handrail in plan; A is the radius point for the quadrant BD; j is where joints will be made. Above the plan, the incline is layed out and the handrail thickness drawn in. A bisection as in number 11 will find a radius point for the easing.

This is not a difficult layout and shaping the rail is straight forward. Though it must be remembered that in all geometrical handrailing, a certain amount of approximation is acceptable.

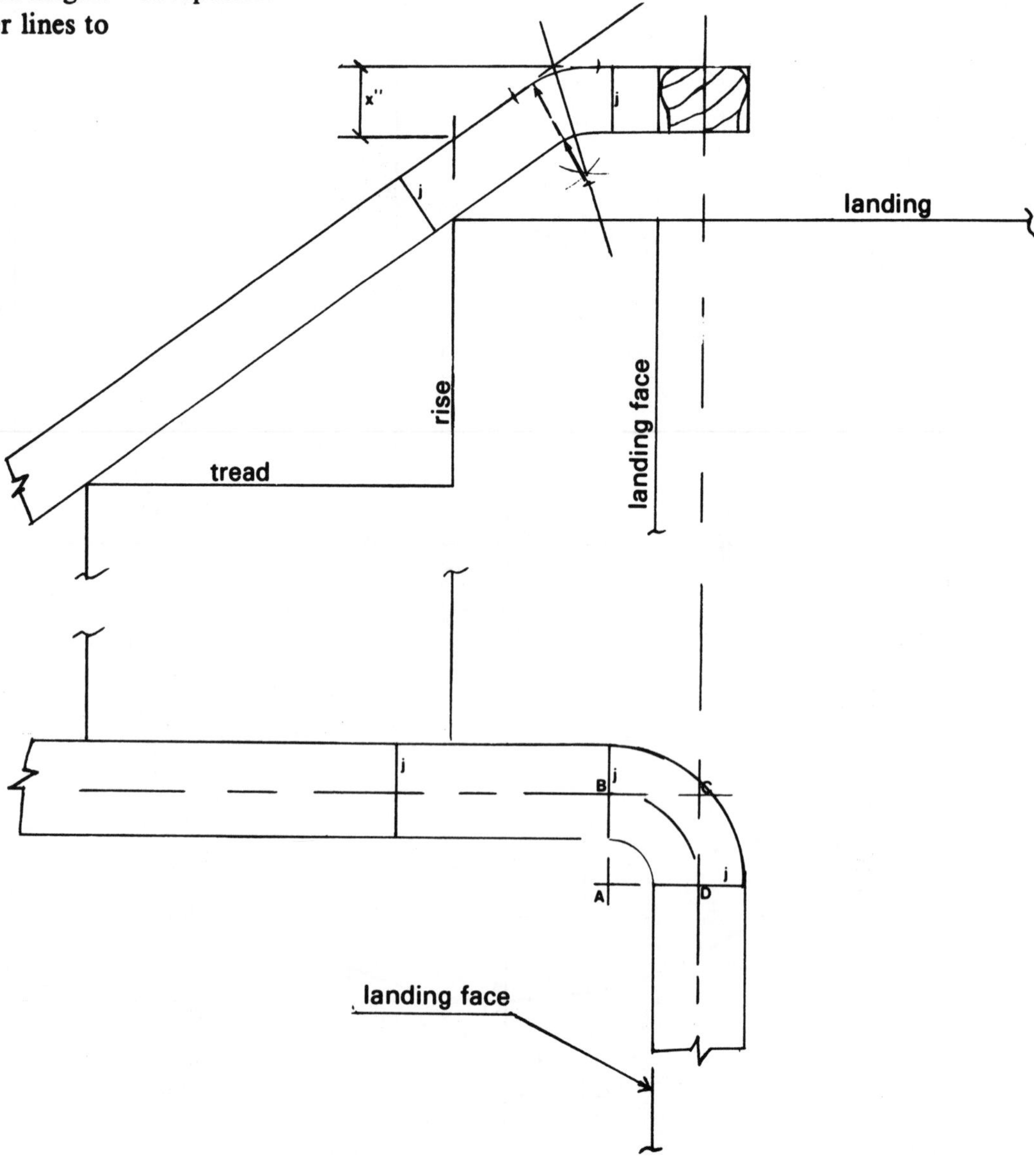

17. If a more continuous flow is desired than the method in number 16, then visualize the rail wrapped about a cylinder. The incline of the rail when viewed at 90° to its top face is elliptical, thus a pattern called a face mould must be made to this elliptical shape:

BCD represent the rail centre lines, A being the radial point of quadrant BD. Above this plan, layout the pitch of the stair and handrail. Project from the plan and layout a rectangular cross section of the rail Y. $A^{I}B^{I}$ equals half the major axis to form the face mould ellipse. To the right of the plan, project from DC and mark off $B^{II}C^{II}$ and $A^{II}D^{II}$. Join B^{II} to D^{II}, then parallels from rail width EF to GH. Complete quarter ellipses using these latter points to form axes. Allow enough straight to make joints j. T represents the thickness of plank required. K shows the action of sliding the face mould to layout the twist. The illustration top right, shows the first layout for bandsawing and below, sliding the template to line up with the pitch line P previously layed out with a T bevel. The handrail section should now be drawn and the piece shaped to finish the "rake to level quarter turn".

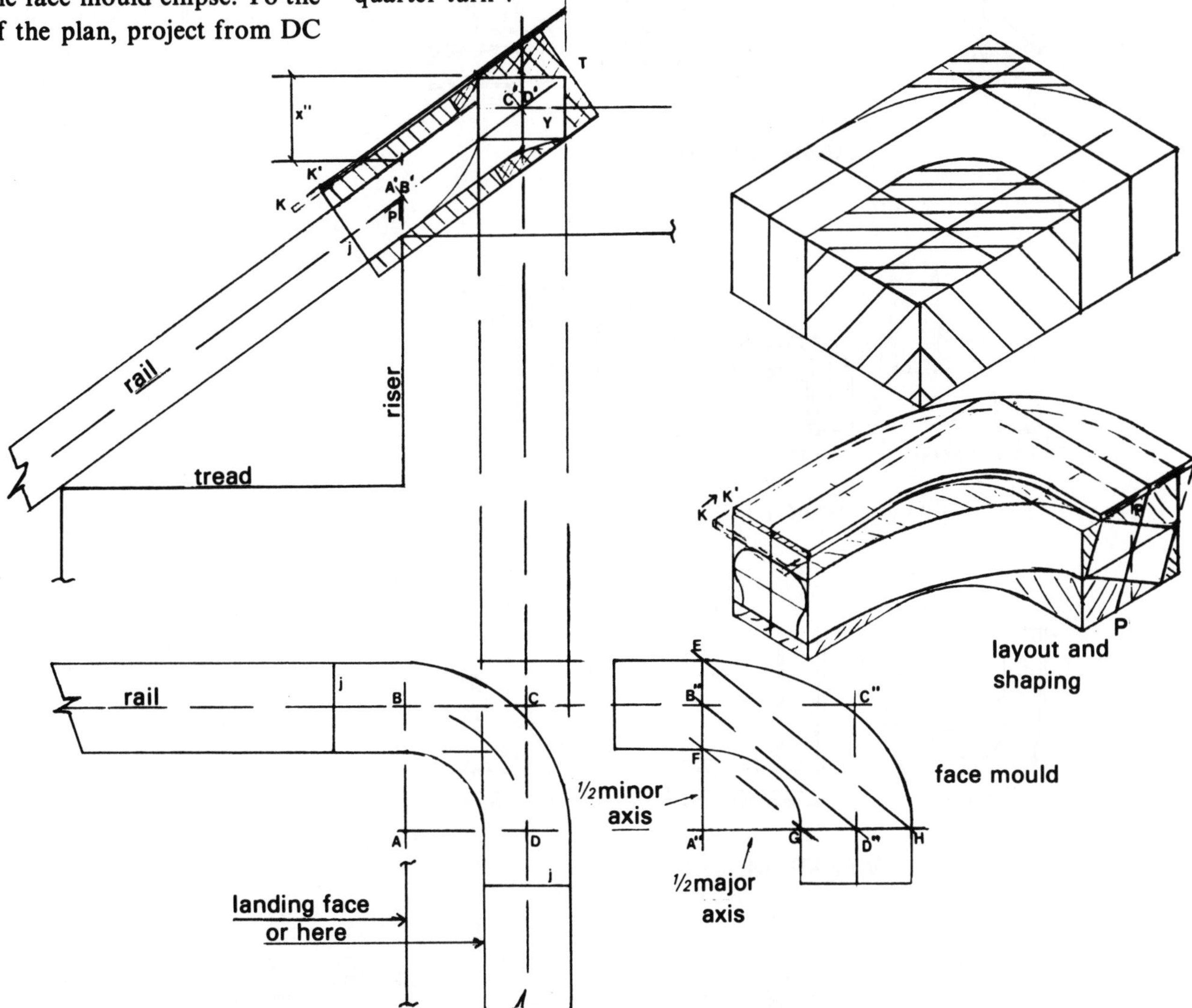

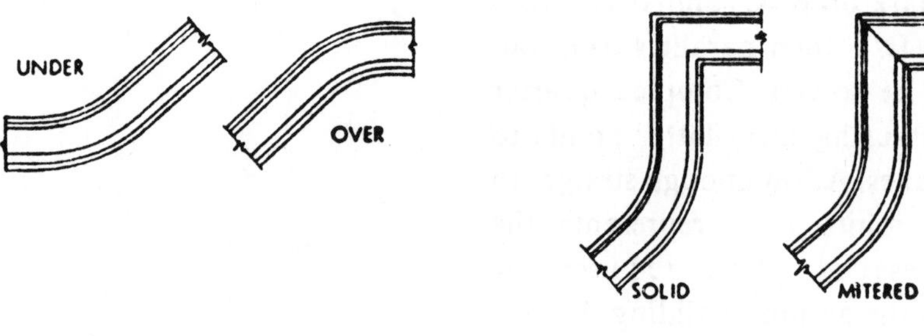

EASEMENT

RAMP OR GOOSENECKS

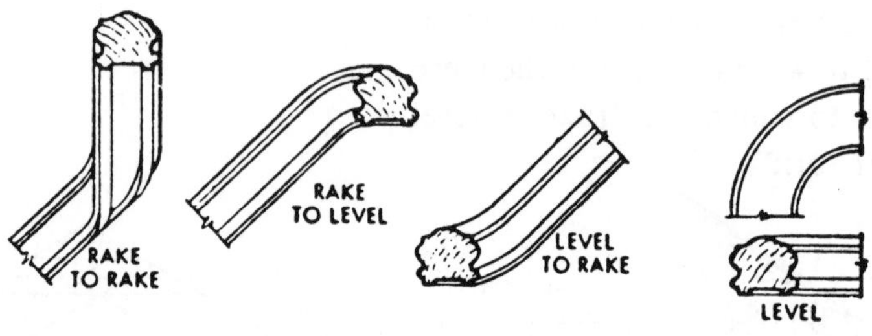

QUARTER TURNS

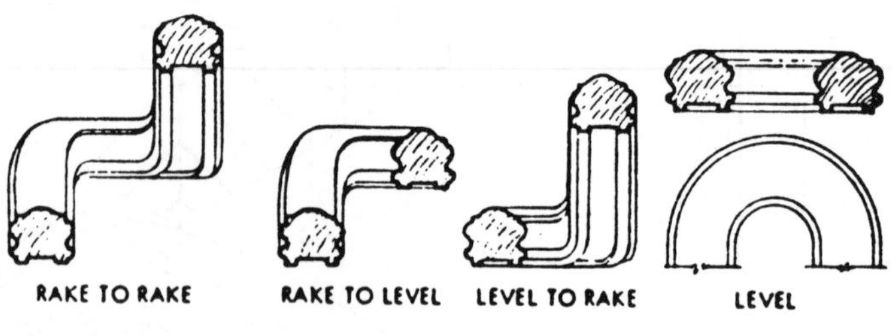

HALF TURNS

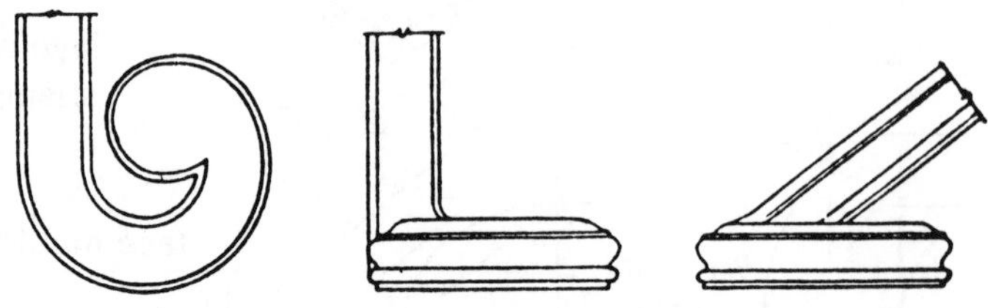

WREATH OR SPIRAL

Discovery Exercises

1. Design a residential stair to suit the following conditions: (single line layout) given a total rise of 105" (2660), a well hole of 122" (3060) long, underside not seen:

Layout:
(a) the two wall strings
(b) the cut string
(c) the newel post
(d) the quarter turn step

select suitable scale

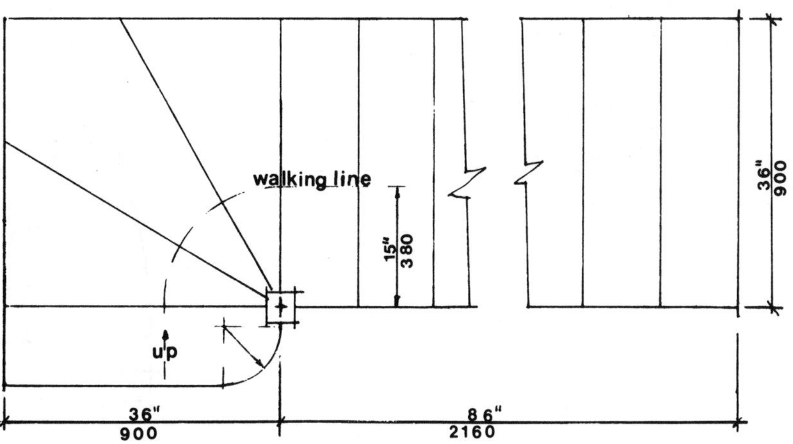

2. Show how the winder tread width in problem 1 might be increased at the walking line.

3. Using a 15" (380) walking line to maintain at least a 9" (230) tread run, draw the plan layout for the circular stair shown. Total rise 105" (2660), single line layout.

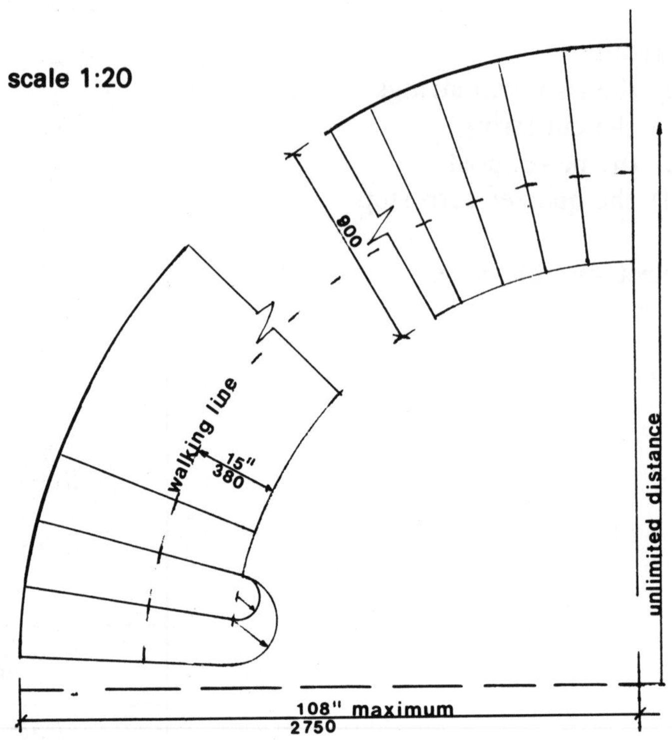

Stairs

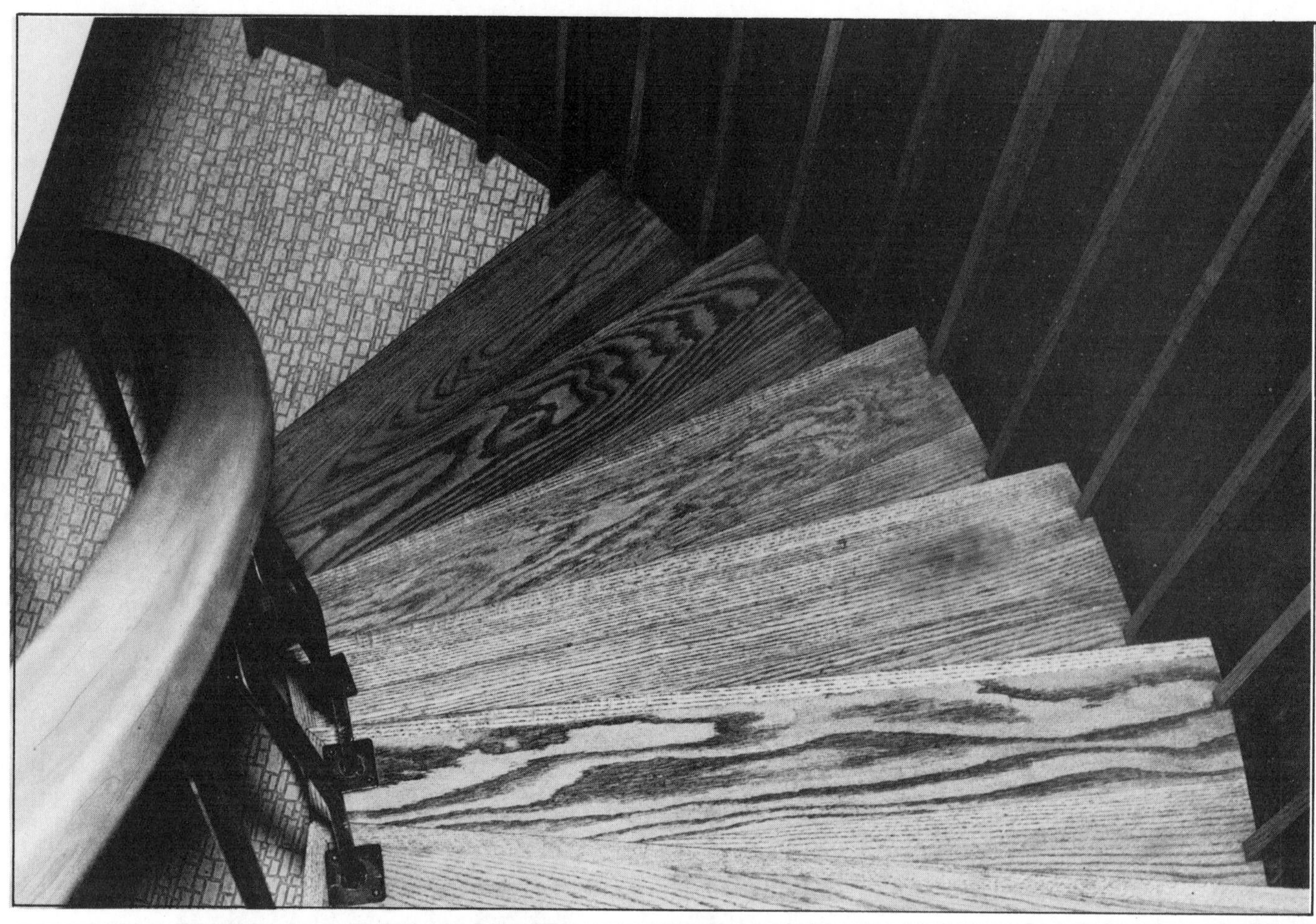

Unit 9
Arches, Tunnels and Vaults

A curve is well-known to structural designers for its support strength and has been used as such for centuries. Hence, we have a carry over of names from the Greek and Roman Classic period. Today however, arches are not only used for support strength, they are used for decorative reasons too.

Wood arches, often called centres, are used singly or in series as a form to place brick, stone or concrete over. After removal, the building material is left in place as the finished product — an arched doorway, window, fireplace, perhaps a subway roof, a sewer or an industrial furnace flue. Such roof lines are known as vaults and tunnels. They may be straight, curved or have several branches.

A carpenter usually has the job of developing, laying out, constructing and setting up such arch shapes. Economics plays an important role in their construction. If only one or two arch centres are needed and the size falls within a standard sheet of ⅝" or ¾" plywood, then cut the shape from the plywood. If many are required, then it may be worth making them from rippings of ply or solid lumber. If the centres are large, say more than 8 feet, then one has to build them with separate members designed with direction of load in mind.

The segmental arch which is part of a circle is layed out from a given span AB and height CD. Layout AB, at its centre, draw a perpendicular, measure height CD. Bisect CB, where this bisection meets the extended CD at E, is the centre to scribe arc ACB. Notice the brickwork is spaced from the key brick down to springing points A and B. Brick joints should be a normal to the curve of any type of arch.

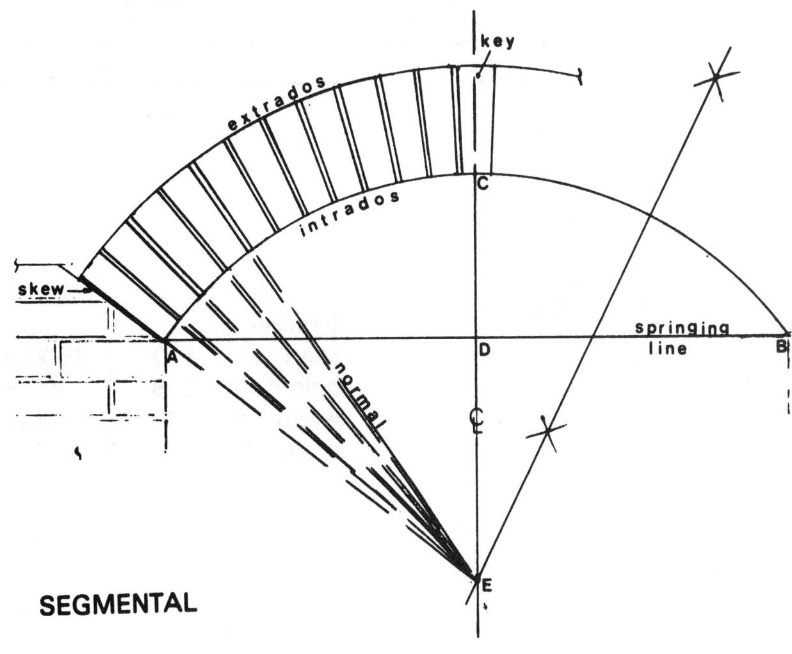

SEGMENTAL

1. To build a 10'-0" x 2'8" centre to form an archway in an 8" wall:

First layout the span and height on the shop floor or on sheets of plywood. Lay down suitable boards as at XY and strike the curve. Reduce the radius to allow for the thickness of lagging, then transfer the joint lines on to the board. Cut 4 face ribs identical and two similar for back ribs, (do not worry about joints) the back ribs scab the members together. Cut two ties and nail two separate frames together. Now fit in the strut braces. Stand the two frames apart, secure them and nail on lagging from the bottom up. This will allow to split up the spacing difference top centre. Nail a spreader at the middle to take out any whip. The lagging should be covered with a plywood skin if concrete is being used.

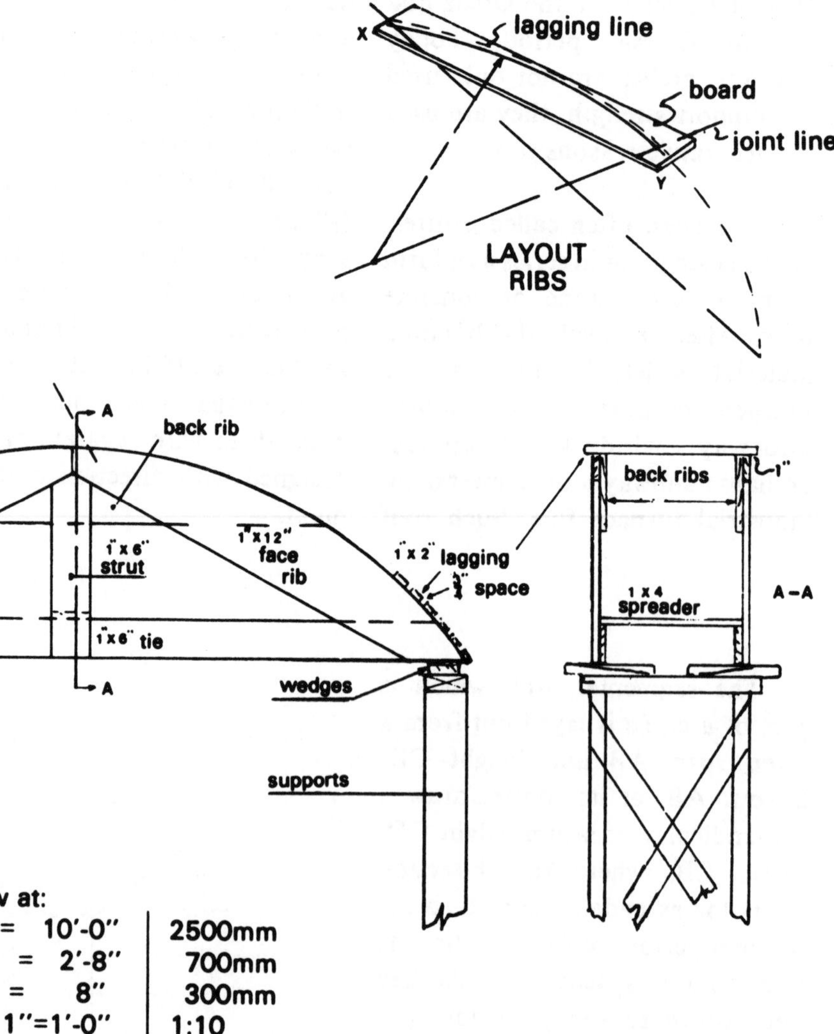

The radius of segmental arches may be found by formula:

$$r = \frac{\tfrac{1}{2}\,\text{span}^2 + \text{height}^2}{2\text{height}}$$

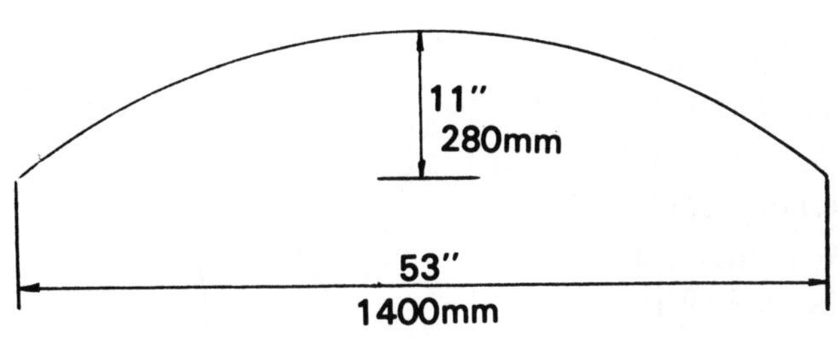

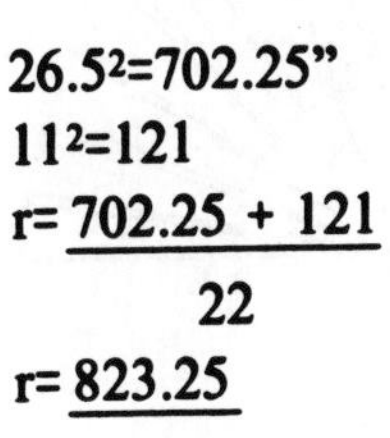

$26.5^2 = 702.25"$

$11^2 = 121$

$$r = \frac{702.25 + 121}{22}$$

$$r = \frac{823.25}{22}$$

$700^2 = 490000$

$280^2 = 78400$

$$r = \frac{490000 + 78400}{560}$$

$$r = \frac{5684000}{560} = 1015\text{mm}$$

$\therefore$ radius = 37.42" or 3' - 1 $\frac{27"}{64}$

$\pm$ 1/16" is a reasonable tolerance

2. The three centered ellipse uses a span AB and a height CD. Draw AB then at its centre, lay off perpendicular CD extended above and below. Join D to A. With radius CA and centre C, scribe to E. With radius DE and centre D, scribe down to F. Bisect AF to give points G and H. With centre H and radius HD, draw arc JDK. Next, use centres G and L and radius AG, to make arcs AJ and KB.

A centre of 8' or less may be cut from one board, but larger ones require rib structure. The one shown would be ideal up to 12' from 1" stock after which more ribs and struts would be necessary.

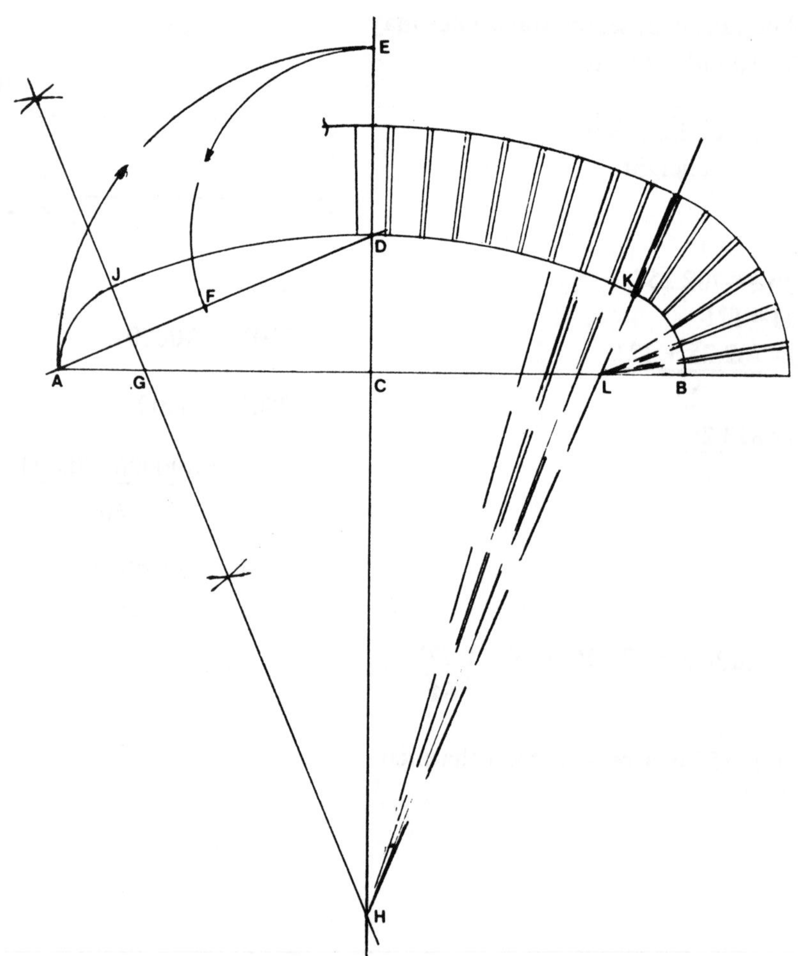

THREE CENTERED ELLIPSE

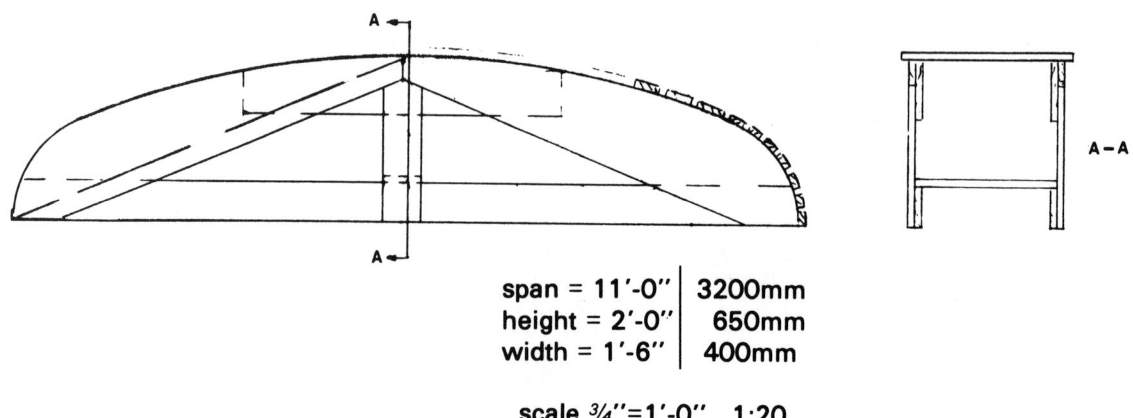

span = 11'-0" | 3200mm
height = 2'-0" | 650mm
width = 1'-6" | 400mm

scale ¾"=1'-0" 1:20

3. Large arches for industrial and commercial ventures should be designed by an engineer because of the tremendous load and direction of pressure:

In large tunnel construction that has a long run, several arches would likely be fabricated from steel, connected as a unit, and hydraulically highered, lowered and traversed on tracks. Quite often short runs are more economically built from wood and changes in direction are usually constructed of wood.

TYPICAL ARCH CONSTRUCTION IN EXCESS OF 30'-0"
METHOD BB OPTION TO AA

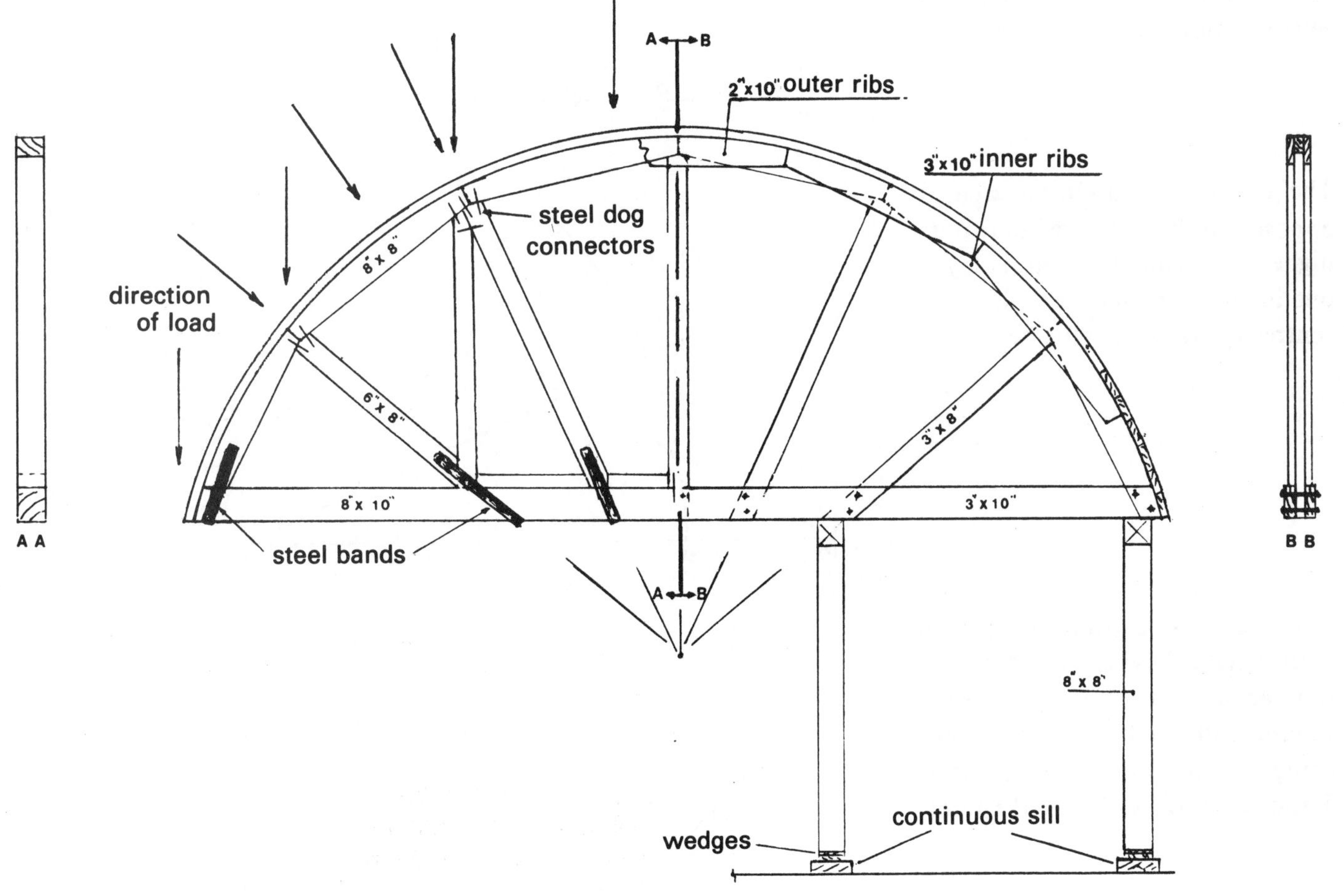

4. The equilateral Gothic is formed over an equilateral triangle. Therefore MN, MO, NO are the same dimension. Use span MN as a radius and M and N as centres.

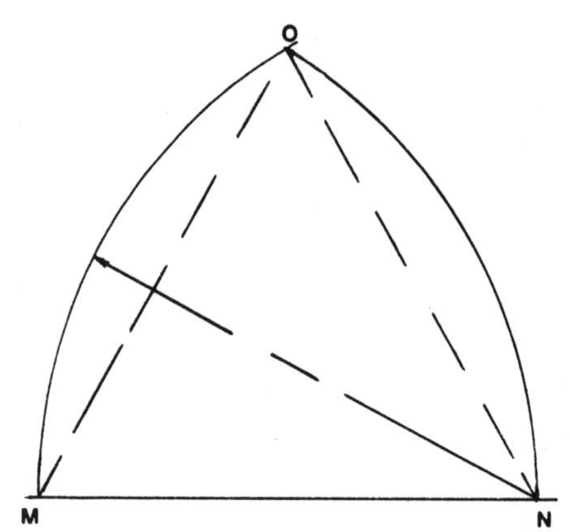

The depressed Gothic, often called a drop arch, is formed about a span MN and a height OP. Bisect NP to find centre Q on the springing line. With radius QN, strike arc NP; reverse from R for arch MP.

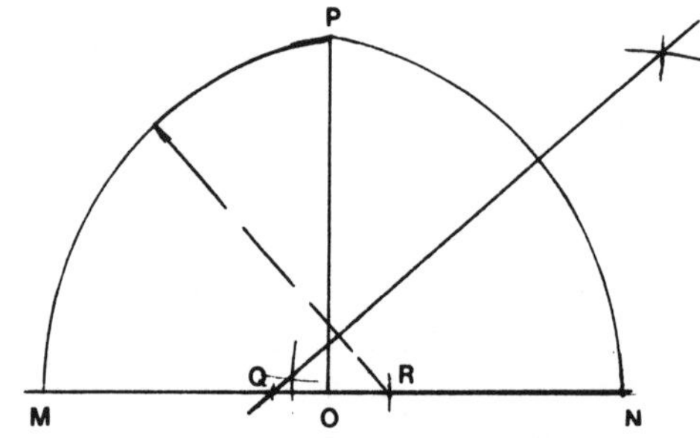

The Lancet, also a Gothic arch, is opposite to the depressed due to its higher rise than span, but is layed out in the same manner. Note the centres Q and R lie outside.

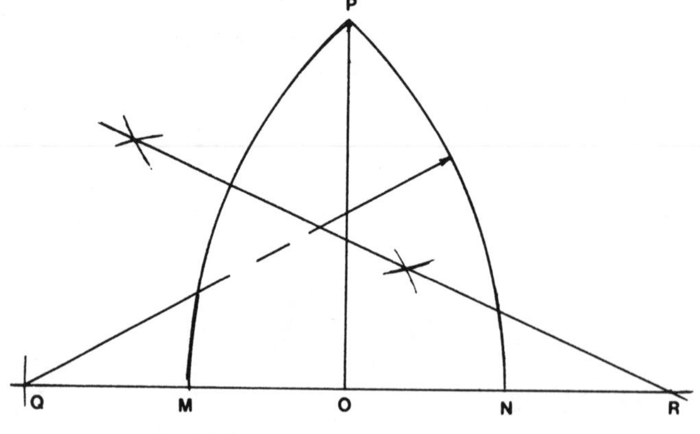

The Tudor has a given span but no given height. Divide span MN into four equal parts as OPQ, make a square with OQSR, each corner being a centre. Radius OM stops at T and radius RT scribes to the point.

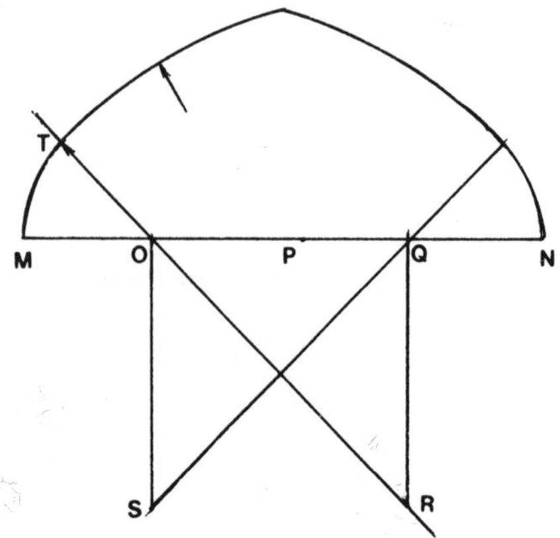

Other Arches

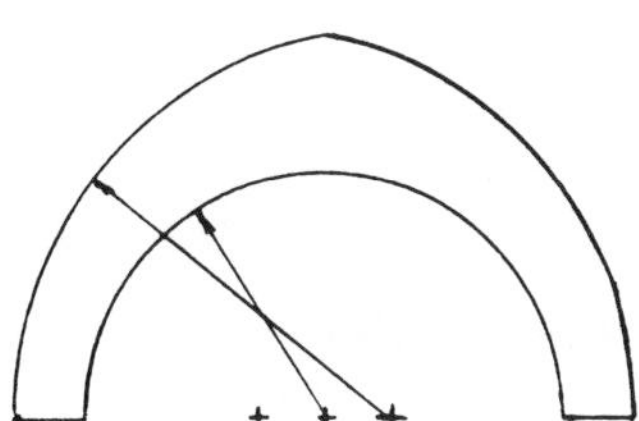

Florentine

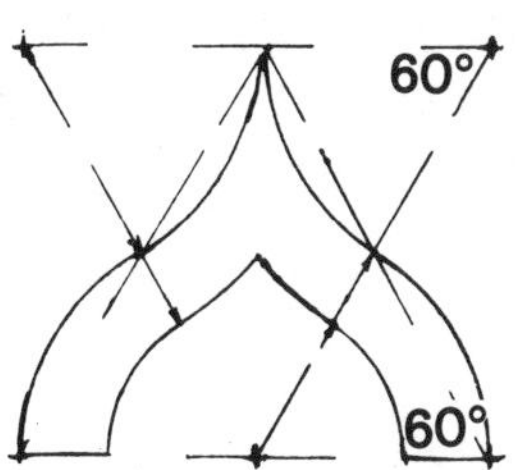

Ogee

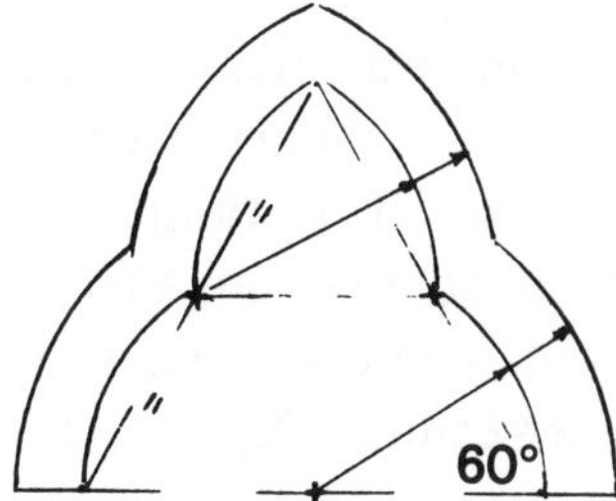

Trefoil

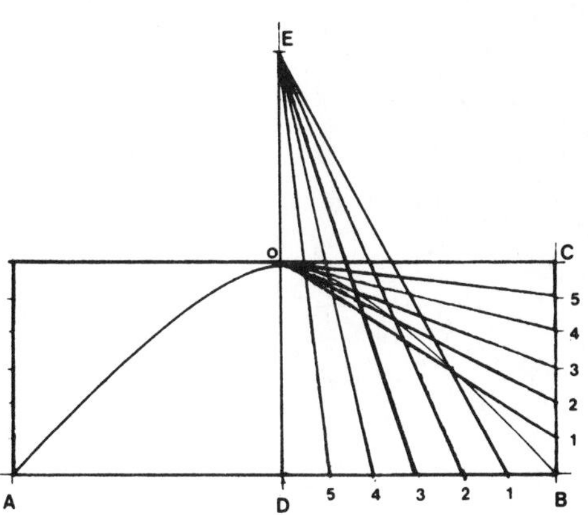

Hyperbola

E is given or may be 2(OD)

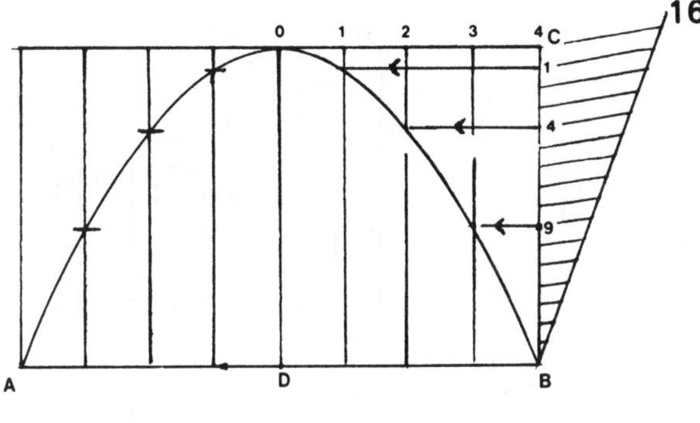

Parabola

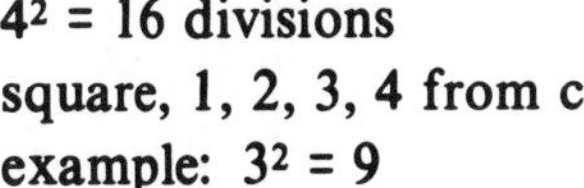

4^2 = 16 divisions
square, 1, 2, 3, 4 from c
example: 3^2 = 9

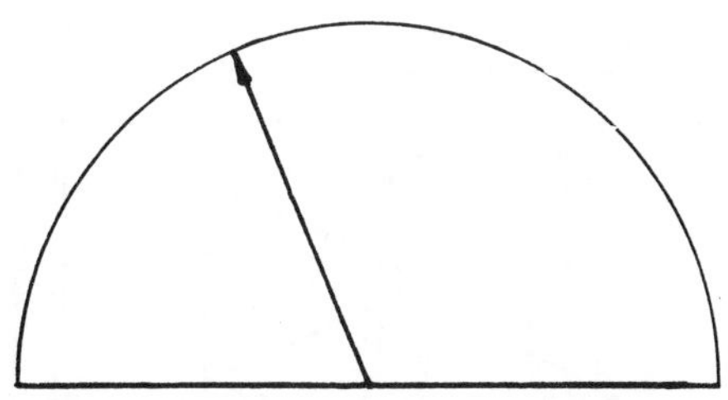

Semicircular or Roman Arch

5. Here is a typical setup for tunnel work. A series of ribs of the desired shape set up on runners supported by well-braced shores, which are sitting on wedges for height adjustment. If a tunnel changes direction, but does not alter in span, then only one rib changes shape at the intersection. Dependent on the centre spacing used, one or more partial ribs may be needed to carry the long point of the sheathing. These are found by projecting from the plan view to the elevation of the original rib shape.

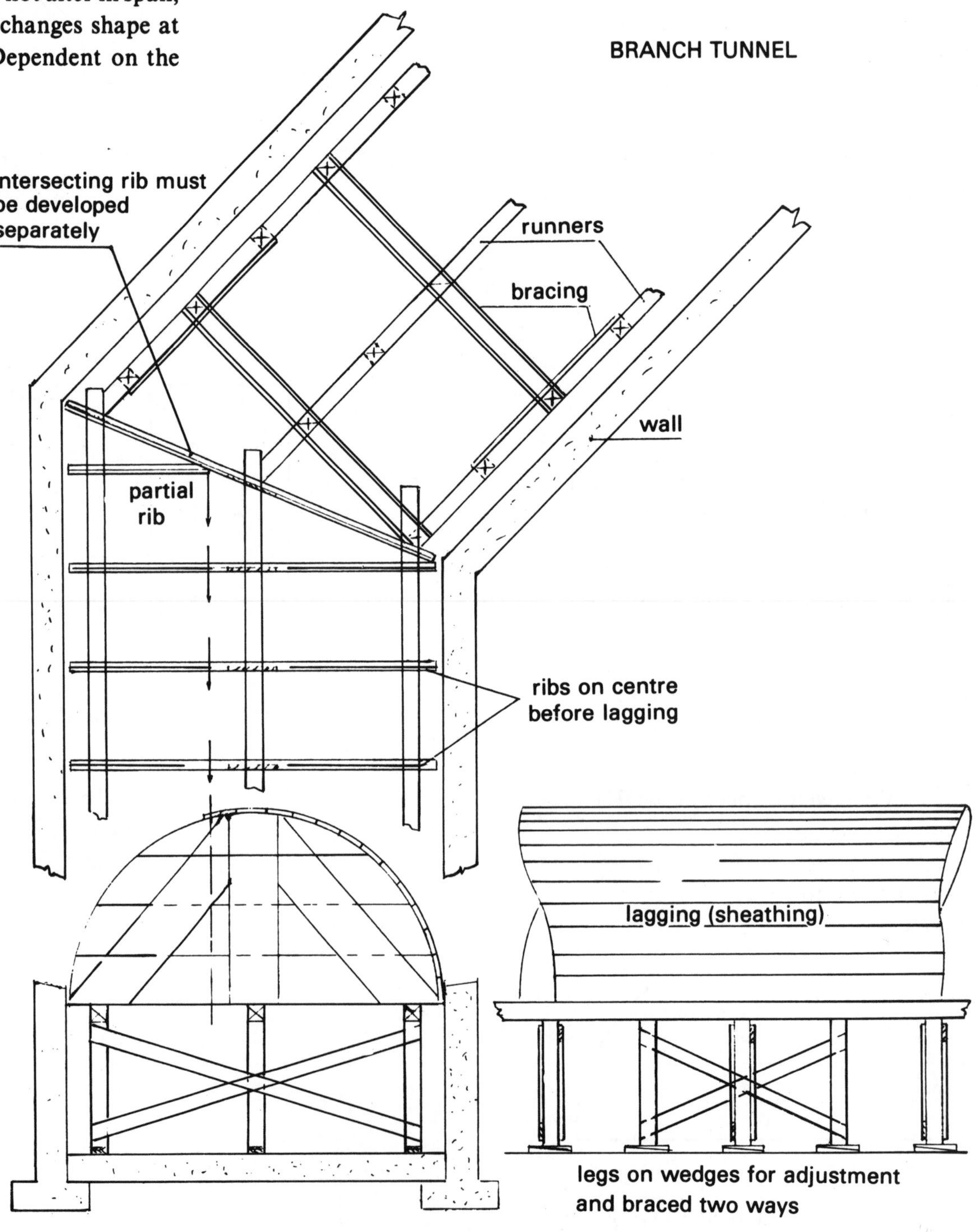

6. To develop a new arch shape, but maintain the same ceiling height at any given point. In this case, the branch changes span, so the intersection is squared off with the greater span. The ribs in the branch all take on the new shape. The procedure for developing the new shape in number 5 is the same as in this problem:

Layout the shape of first given arch on the plan, in this case semicircular, cut the span into equal parts as 0-10. Project these points to the intersection, then continue the lines parallel with the branch. Pick up height 1-1^{1} and lay off at 1-1^{11}, and 2-2^{1} to 2-2^{11} and so on. Now using curves or freehand, complete the new shape.

If sheet material is being used for sheathing, its shape must be developed by unfolding it to the side. From the intersecting span at 10, lay off a stretch out line. With compasses, pick up the distance over the original rib 10-9^{1} and transfer to the stretch out, then 9^{1}-8^{1} and 8-7^{1} etc. Now project lines up from the new points to meet horizontal lines from the intersection. Complete a curve where these lines meet. This curve will fit to the square ended branch.

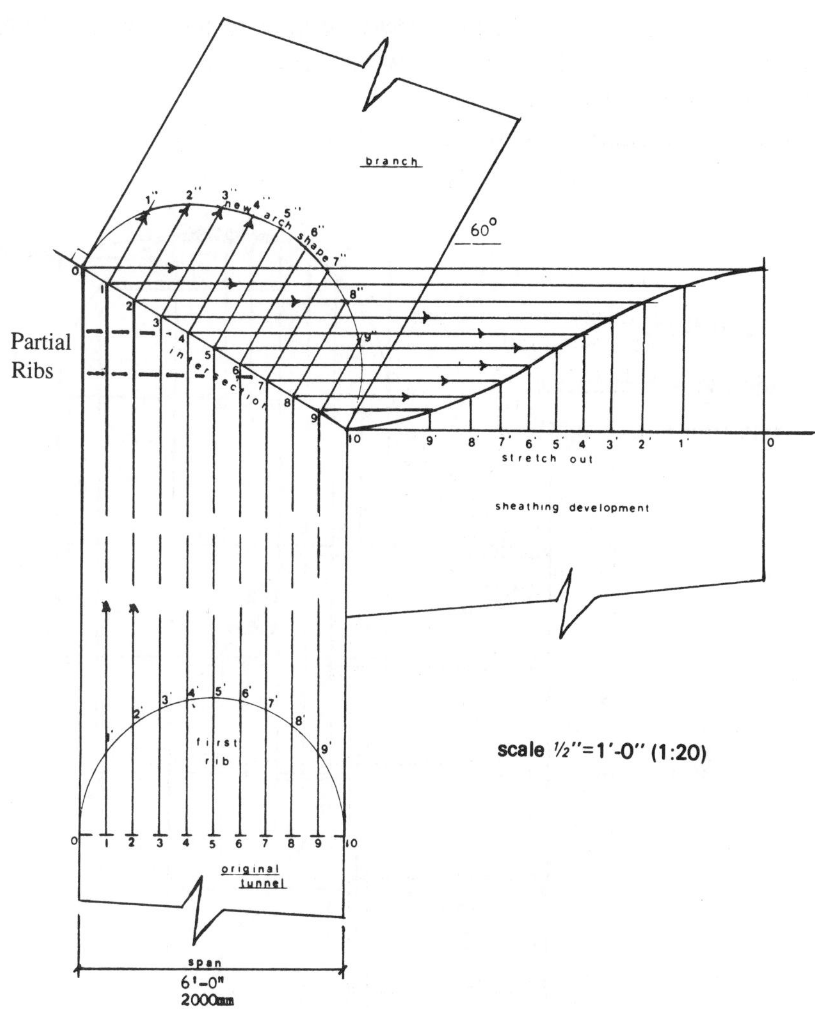

7. When one tunnel crosses through another with the same roof height a new rib shape must be found as in number 6:

In forming up, let the ribs with a wider span run through and the other tunnel overlap. The development procedure for ribs and sheathing is the same as in previous work.

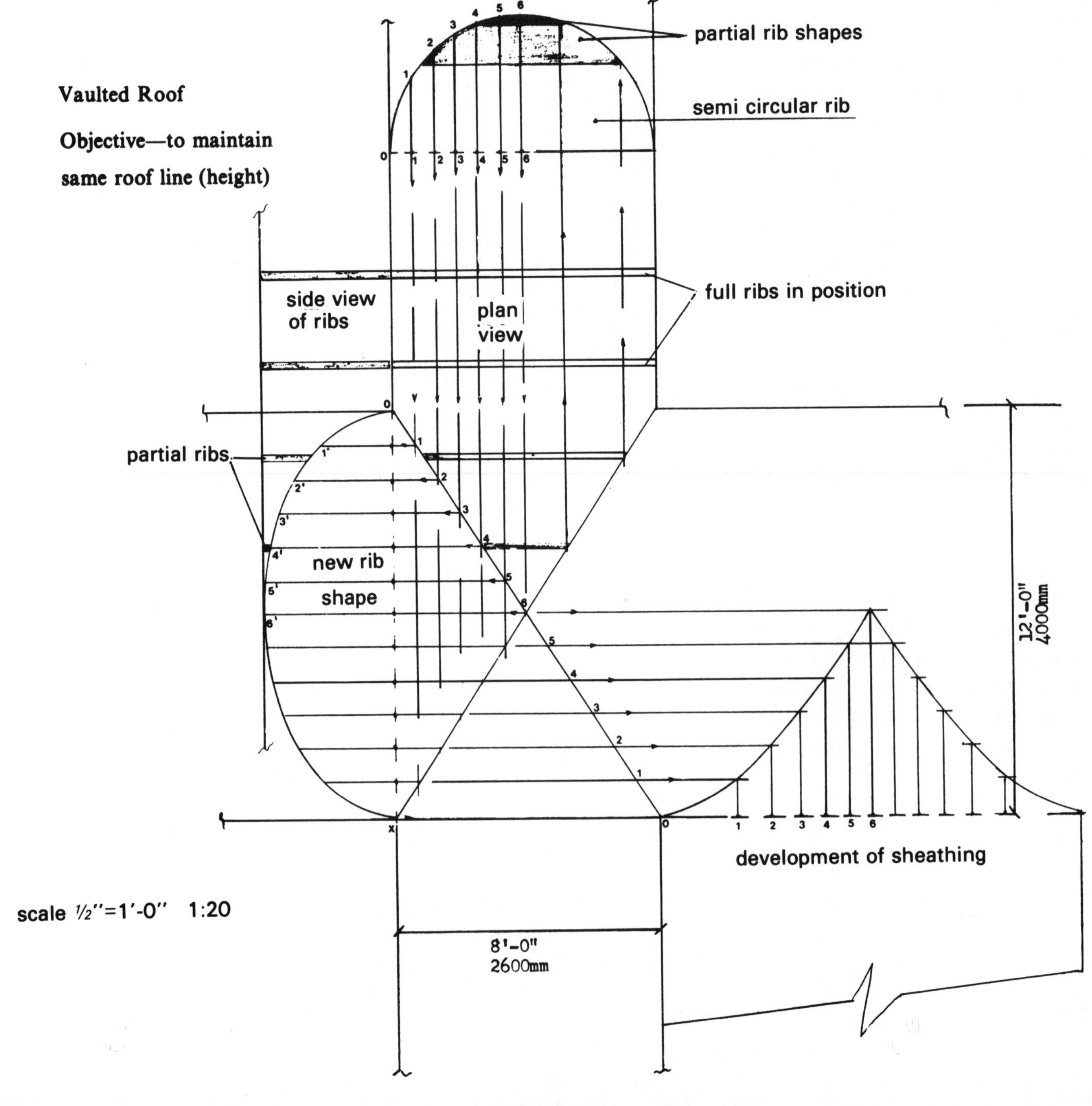

Discovery Exercises

1. Layout and design a suitable arch centre to carry brick. (Use more than two face ribs.)

style — segmental
span = 11'-0" 3400
height = 2'-9" 850
wall thickness = 9" 300
scale ¾" = 1' 1:20

Label all parts and dimensions.

2. A tunnel is to built with a semicircular vaulted roof.

a. Layout the rib shape of the intersecting vault maintaining the same roof height.
b. Develop the shape of sheathing to cover surface A. Scale ½"=1' (1:20)

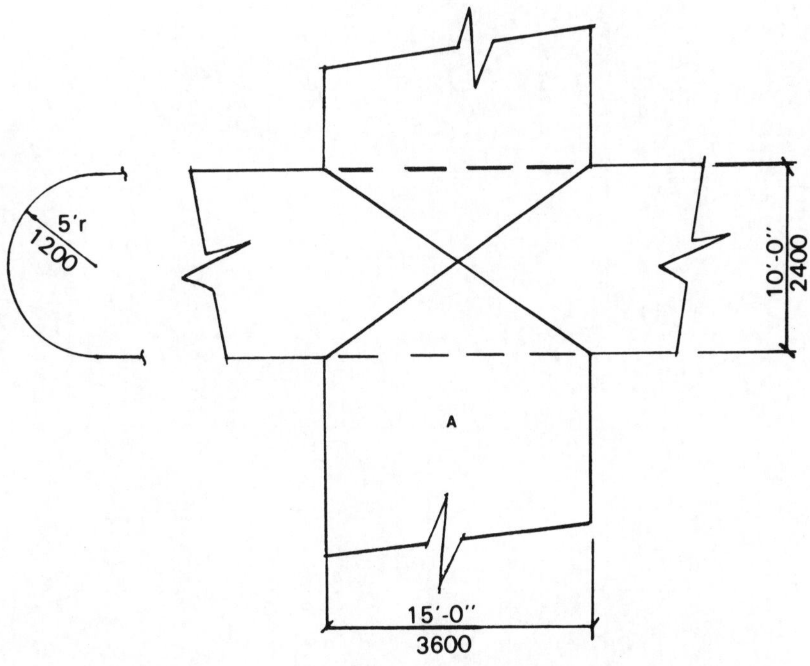

Courtesy of George Brown College —
H. Freund, master brick and stonemason.

Unit 10
Roof Problems

It is assumed here that the reader has a reasonable understanding of roof framing. As in all development work, one must attempt to visualize the unfolding of one plane on to a horizontal plane so that true shapes are exposed in a plan view.

1. The diagram shows a relationship to the right angle; a steel square being used for 90°. A student of this geometry should try to relate to a carpenter's framing square, insomuch as lines (total dimensions) are used to form various bevels and what dimensions would be used on a square to obtain the same.

Study the common rafter OP. P is used as a hinge, with P as centre and point length OP as radius, the rafter swings over to the horizontal, then produced to the plan. Now the rafter can be seen unfolded in a plan view.

NOTE: O¹P¹ is still the point length where NP¹ is the run of roof. We have not changed any measurement, just the view.

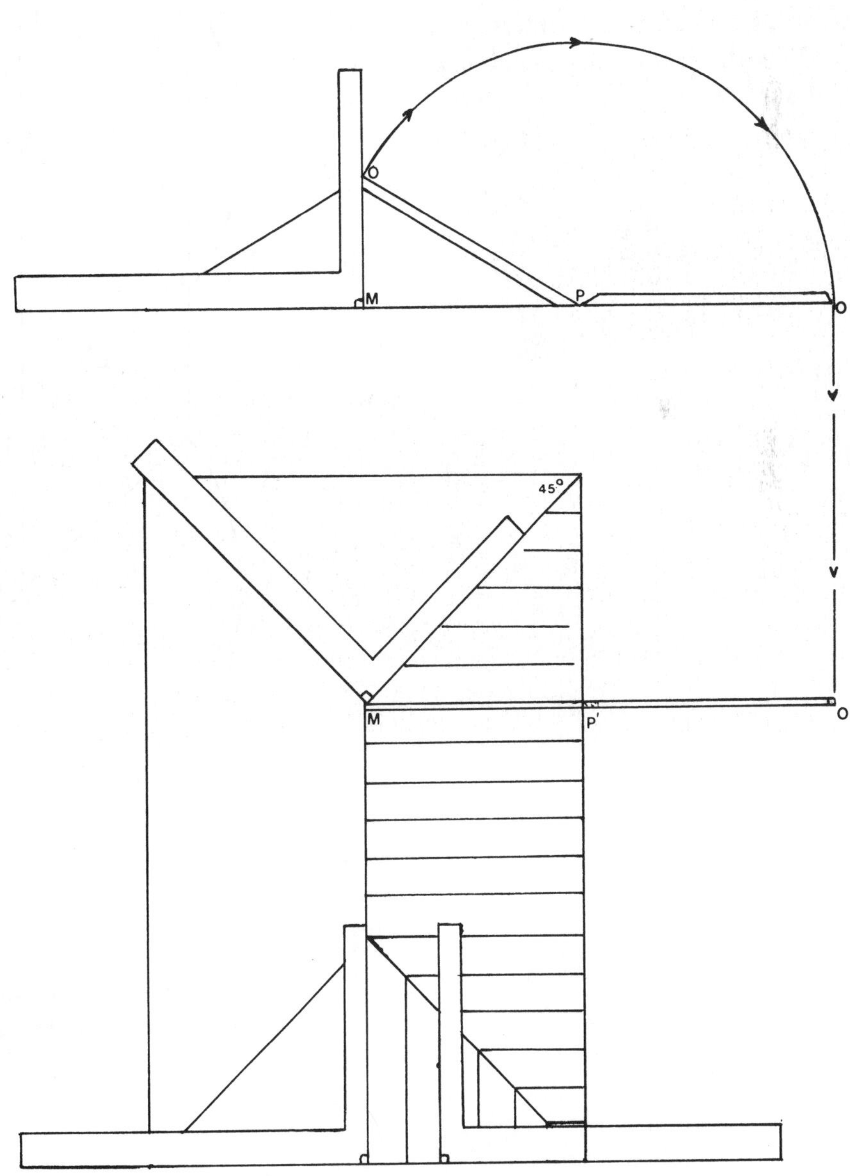

2. A simple hipped roof is shown here with all roof surfaces developed to expose their true shape. Eave projection is not considered.

Use 1 as a centre and produce common point length 1 - 2 to the plan. Intersection 0 is projected to meet at numbers 3 and 3. 3 and 4 represent the point length of hips. E[1] is the true shape of that side of the roof. To find D, use 4 as a centre, 4-3 as radius, and produce to intersect at 5. D[1] is the true shape of the end roof surface.

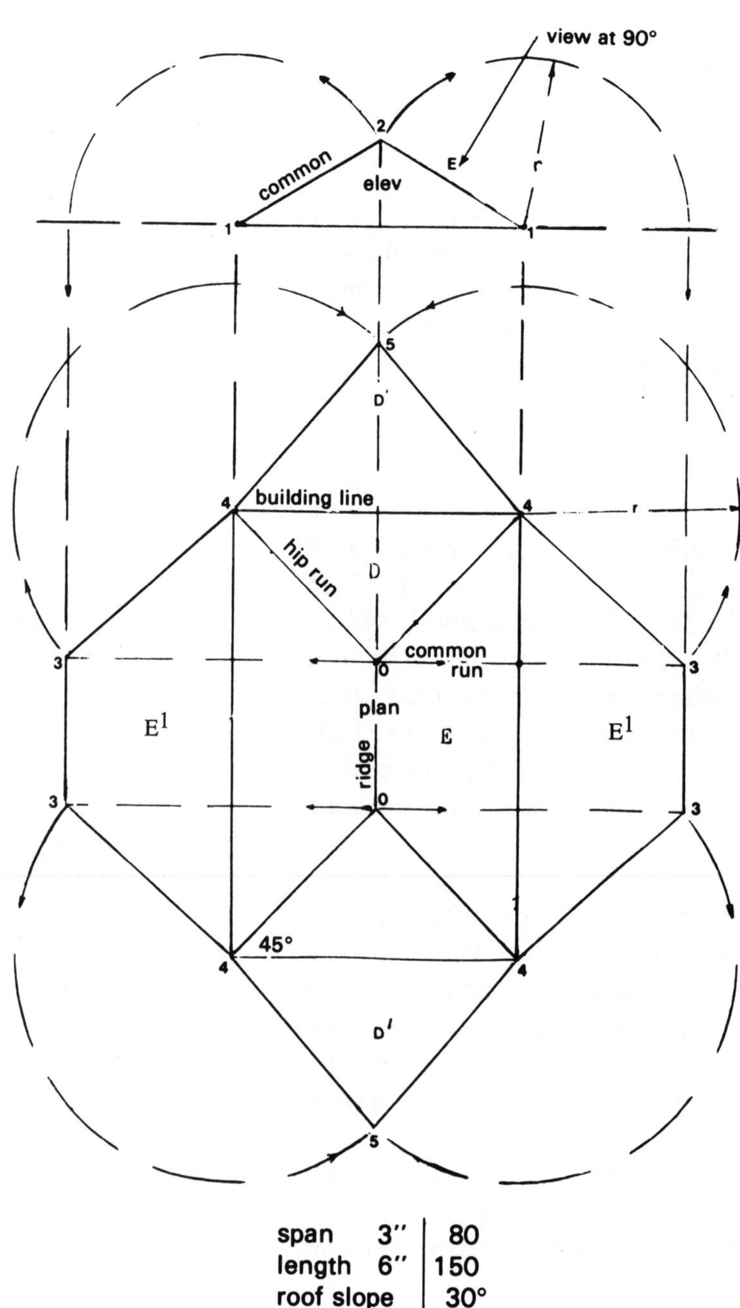

A hip and common rafter may be developed on this plan and used as a cutout model.

Nine Roof Bevels

3. When the equal pitch roof bevels are well understood, it will be appreciated that all the bevels can be drawn on a board in about 15 minutes. It must also be remembered than once drawn, the bevels are there from start to finish of the job, as opposed to the framing square, which must be set and reset. To draw these particular figures, use only one corner of the roof plan and a half vertical section.

Visualize the common rafter in its fixed position on the elevation section.

Visualize the hip as a set square standing on the plan then pushed over.

Using the XY line as the split-layout line:

Commence by setting out the pitch of the roof (end elevation) showing (1) the common rafter level cut, and (2) the common rafter plumb cut. Next, draw the corner of the plan below, develop the plane of the hip and indicate (3) the hip cut, and (4) the hip plumb cut. Set out another diagram. Use the length of the com. as radius and determine (5) the jack side cut. (6) is the sheathing face and purlin edge cut. At right angles to the hip run, draw a line to intersect the Bld. Line; with A as centre, scribe a tangent from P to the run of hip, and form (7) the backing bevel (dihedral). Set out a third diagram. At right angles to the common rafter point length, mark off the total rise to determine (8) the sheathing edge and purlin face cut. Again develop the point length of hip and at right angles to it, mark off the run of hip. This gives (9) the hip side cut.

Now try to picture a framing square laying in each right angle — see what figures are being applied and relate to the rules for cuts by framing square (example: using the twelfth scale.) Total rise and total run give level cut and plumb cut of common rafter. Number nine is using point length of hip and run of hip. Mark on the point length side to lay out hip side out.

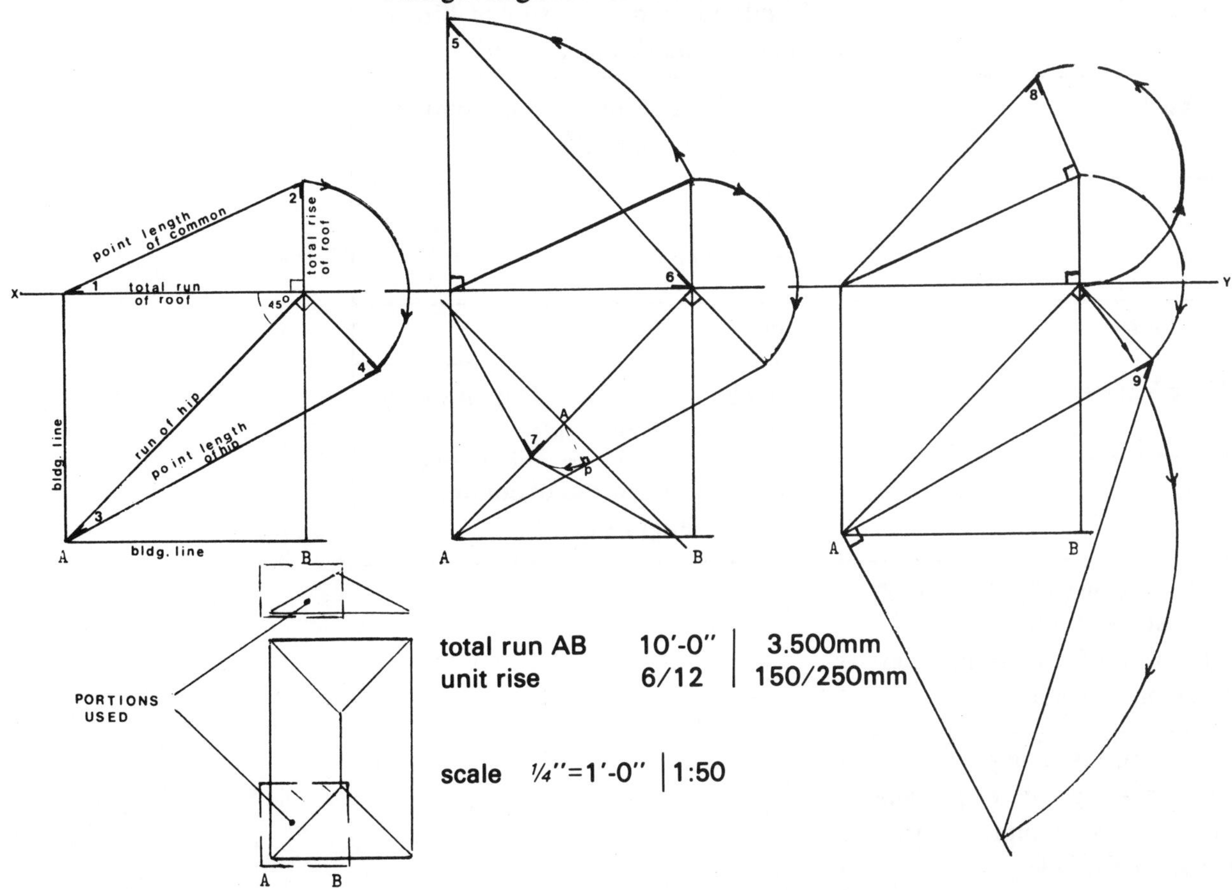

With Unit or Total Measure:
Rules for Framing Square Application. Using 1/12" to 1' scale, or 1:10 on a metric square.

RAFTER	PLUMB CUT
Common	Total rise of roof and total run of common — mark on rise side.
Hip & Valley	Total rise and total run of hip or valley — mark on rise side.
Jacks	Total rise and total run of common — mark on rise side.
	LEVEL CUT
Common	Total rise of roof and total run of common — mark on run side.
Hip & Valley	Total rise and total run of hip or valley — mark on run side.
Jacks	Total rise and total run of common — mark on run side.
	SIDE CUT
Common	Square
Hip & Valley	Point length and run of hip or valley — mark on length side.
Jacks	Point length and run of common — mark on length side.
Sheathing face and purlin edge cut use — point length and total run of common rafter; mark on run side.	
Sheathing edge and purlin face cut use — point length and total rise of common rafter; mark on rise side.	

4. An equal pitch, but oblique ended roof is developed to exemplify the workability of this geometry. Notice the hips, being more acute on one side than the other creates a different side cut, at the bottom 6 and 7, at the top 5 and 8. Jack rafter point length and side cuts are seen on 9. 0 shows each dihedral, which is preferred to the drop method for this type of roof.

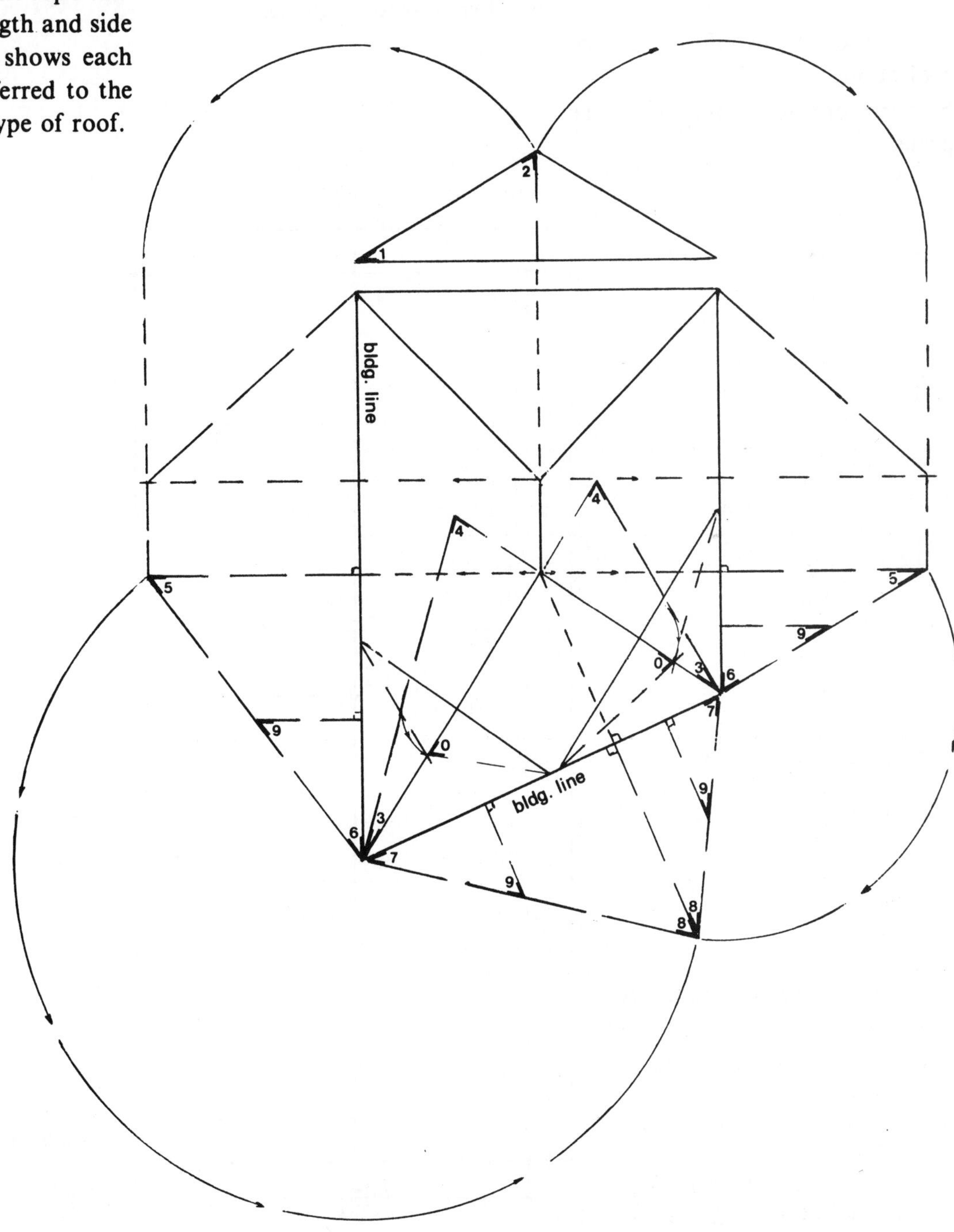

Unequal Pitch Situation

5. To maintain a level soffit and equal projection, the rafters on the minor roof must be raised. To find the amount to raise, superimpose one roof section over the other:

M = run of main roof, 0 = desired projection
N = run of minor roof, 0[1] = same projection

P = soffit line of main roof extended back to intersect 0[1].
From this intersection, R is joined to the ridge. Q shows the amount to raise the rafter plate. The dotted line is rafter R before raising.

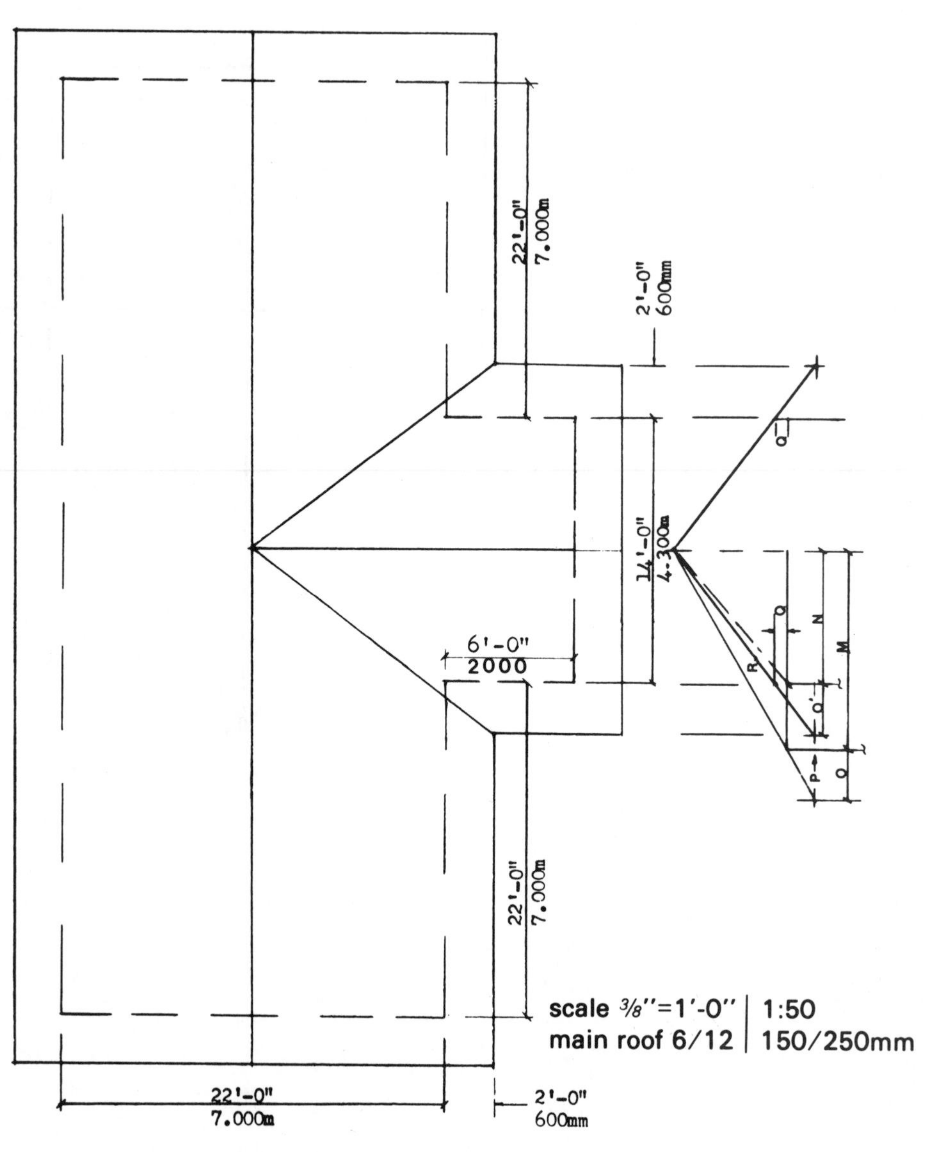

6. To discover the lengths and cuts of rafters in an unequal pitch roof, the members should be layed out geometrically; full size or to scale. 02 identifies the total rise of the roof. 1 is a common rafter level cut, 2 is a common plumb cut. Using the same total rise, develop the valley rafter exposing its point length 3 to 4, with 3 a level cut and 4 the plumb cut. Lay out the hip in the same manner, 5 is the hip level cut, 6 the plumb cut, and 5 to 6 the point length. When projected at 90° from the run U, V, W, being equal, show the amount of drop in the cornice. By developing the roof surfaces, other lengths and bevels may be identified as at 7-8-9. Q is the amount needed to raise the plates on the minor roof.

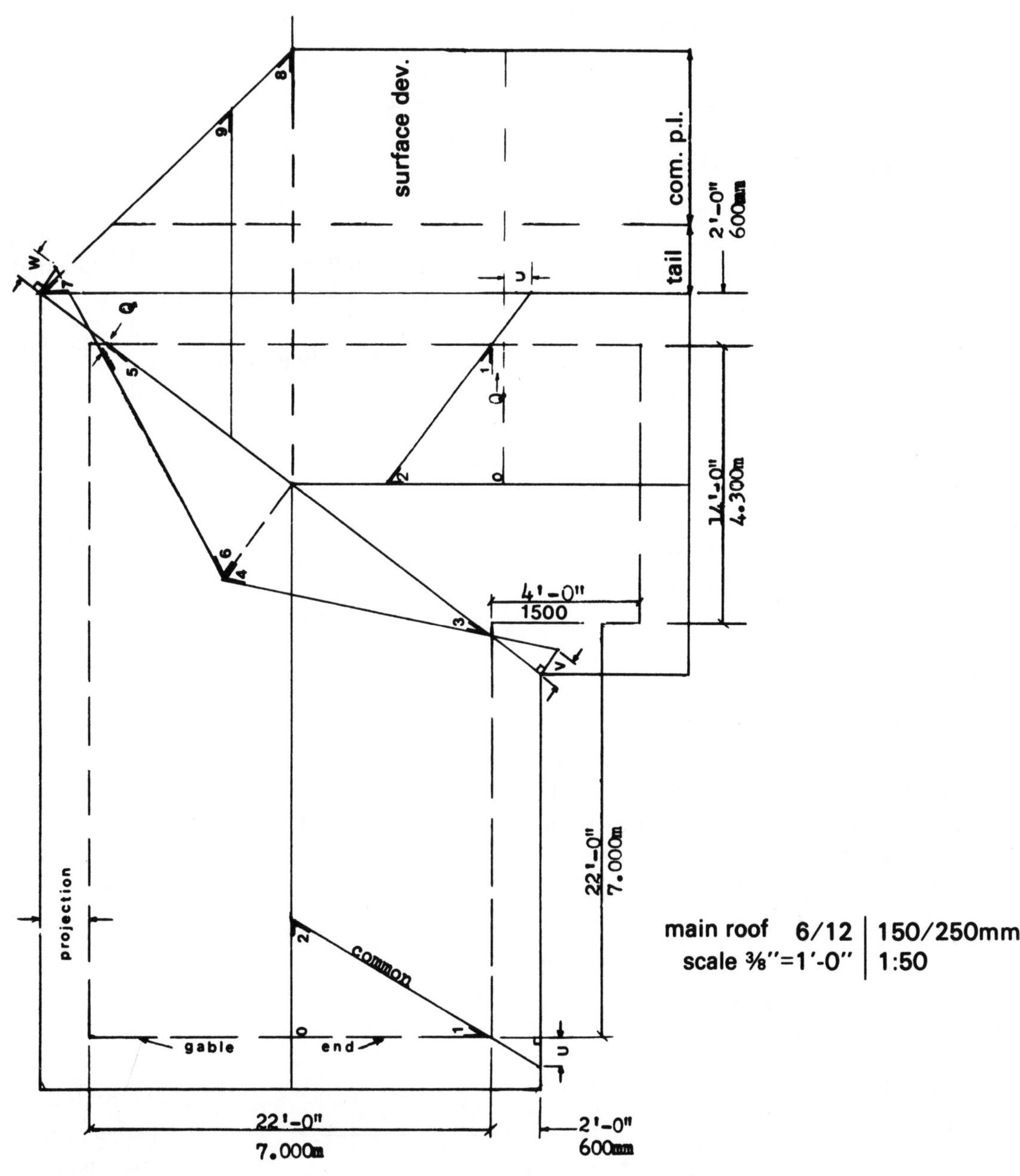

Raising Wall Plates, problems 5-6.

An intersecting roof of unequal pitch requires special treatment to maintain a level soffit and equal projection. The minor roof wall plate may have to be raised.

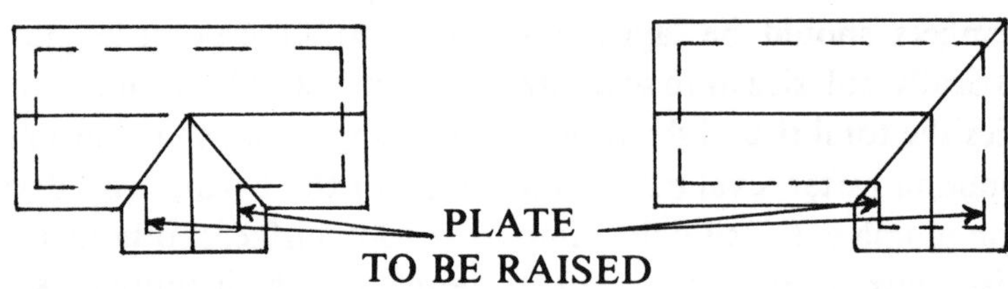

The amount to raise the plate may be found by a full size or scale layout, or by math.

Required Information:

Main roof run	11' — 0"
Minor roof run	7' — 0"
Projection run	2' — 0"
Main roof slope	6/12
Minor roof slope	Unknown

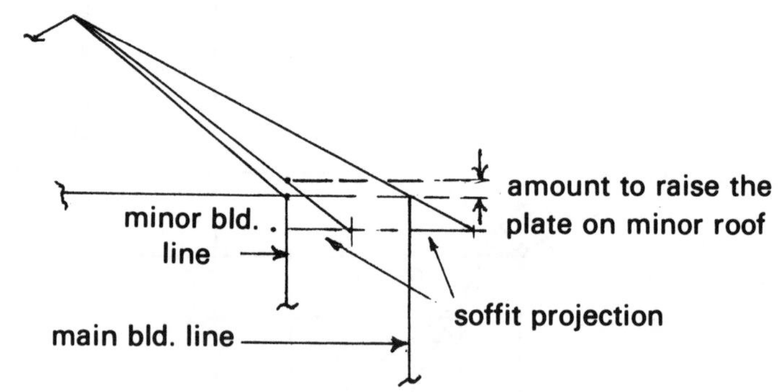

Ridges are level

Extra height of minor plate — difference between main and minor cornice rise.

Mathematical Procedure:

a. Total rise of main - - - 11 x 6" = 5'-6"
b. Total rise of cornice - - - 2 x 6" = 1'-0"
c. Total overall rise: roof + cornice = 5'-6" + 1'-0" = 6'-6"
d. Total overall run of minor = run + cornice projection = 7'-0" + 2'-0" = 9'-0"
e. A:B::C:X

   9:6.5::2:X

   9X=13

   X=13/9

   X=1.444 or 1'-5 5/16"

∴ 1'-5 5/16" minor cornice
−1'-0" major cornice
=0'-5 5/16"

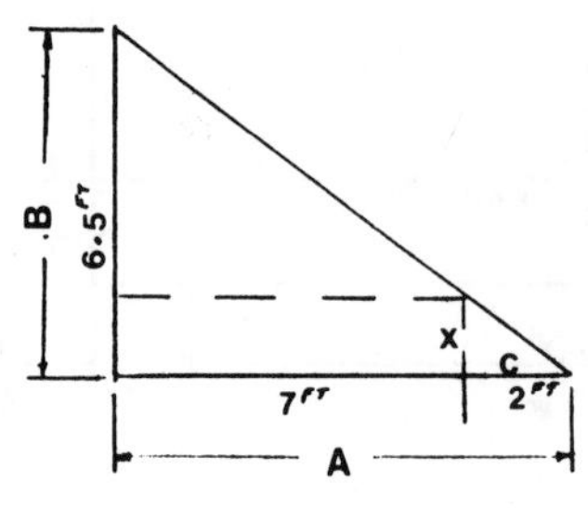

To maintain a level soffit and equal projection, the minor rafter plate shall be raised 5 5/16".

Hip and Valley Offset, for 5-6.

Valley Offset = Projection minus X

A:B::C:X
13:9::2:X
13X=18
X=18/13
X=1.384 or 1'-4⅝"
Offset=7⅜"

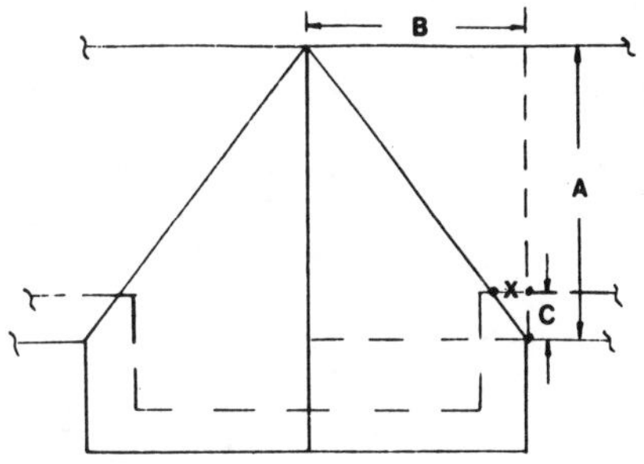

Hip Offset = X minus Projection

A:B::C:X
9:13::2:X
9X=26
X=26/9
X=2.888 or 2'-10 11/16"
Offset = 10 11/16"

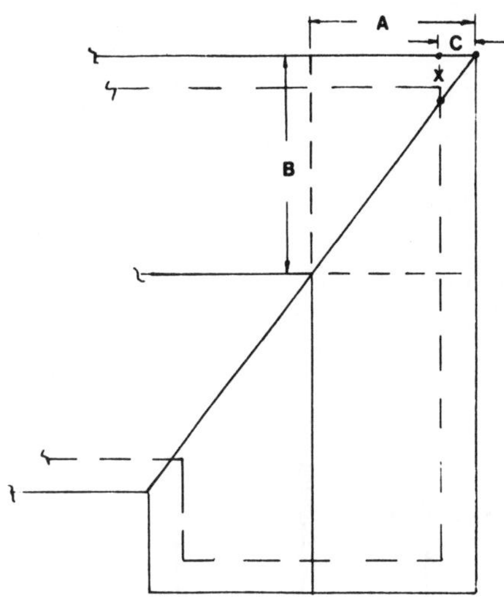

Using the metric dimensions:
the plates would have to be raised 177mm
the valley offset would be 198mm
the hip offset would be 295mm

Roof Problems

The geometry for unequal pitch hip and valley side cuts as layed out with actual stock size, over a scale plan.

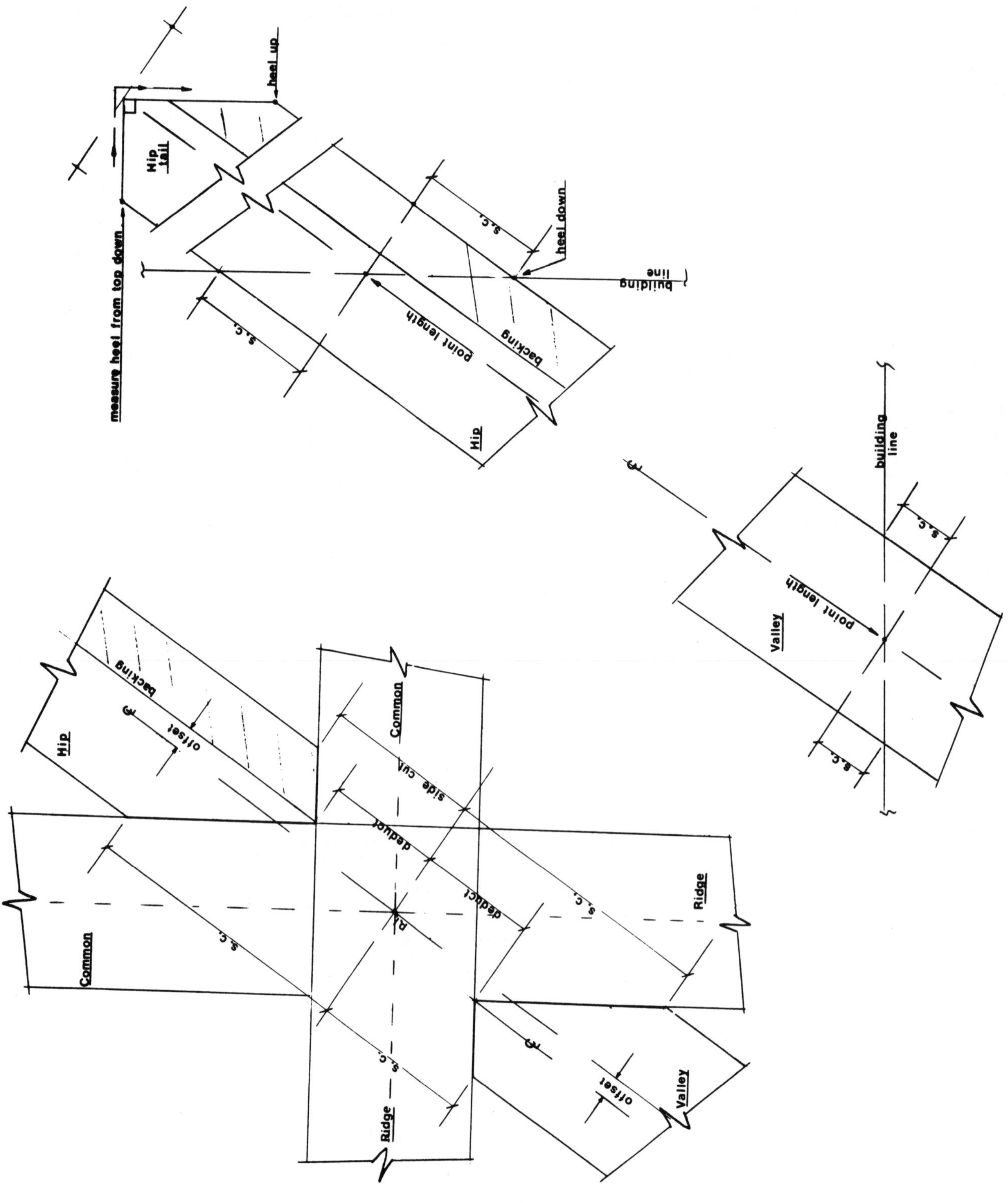

Unequal hip and valley side cuts are better found by drawing full size and transferring measurements to material; across a level line then plumb to the edge. The hip heel is arrived at by measuring the common heel at the tail side cut, down the short side, square, around the previously made side cut—then up the other side—this will determine the point for backing. The heel can now be measured at the side cut line down from the backing-levelled and squared under to the first side.

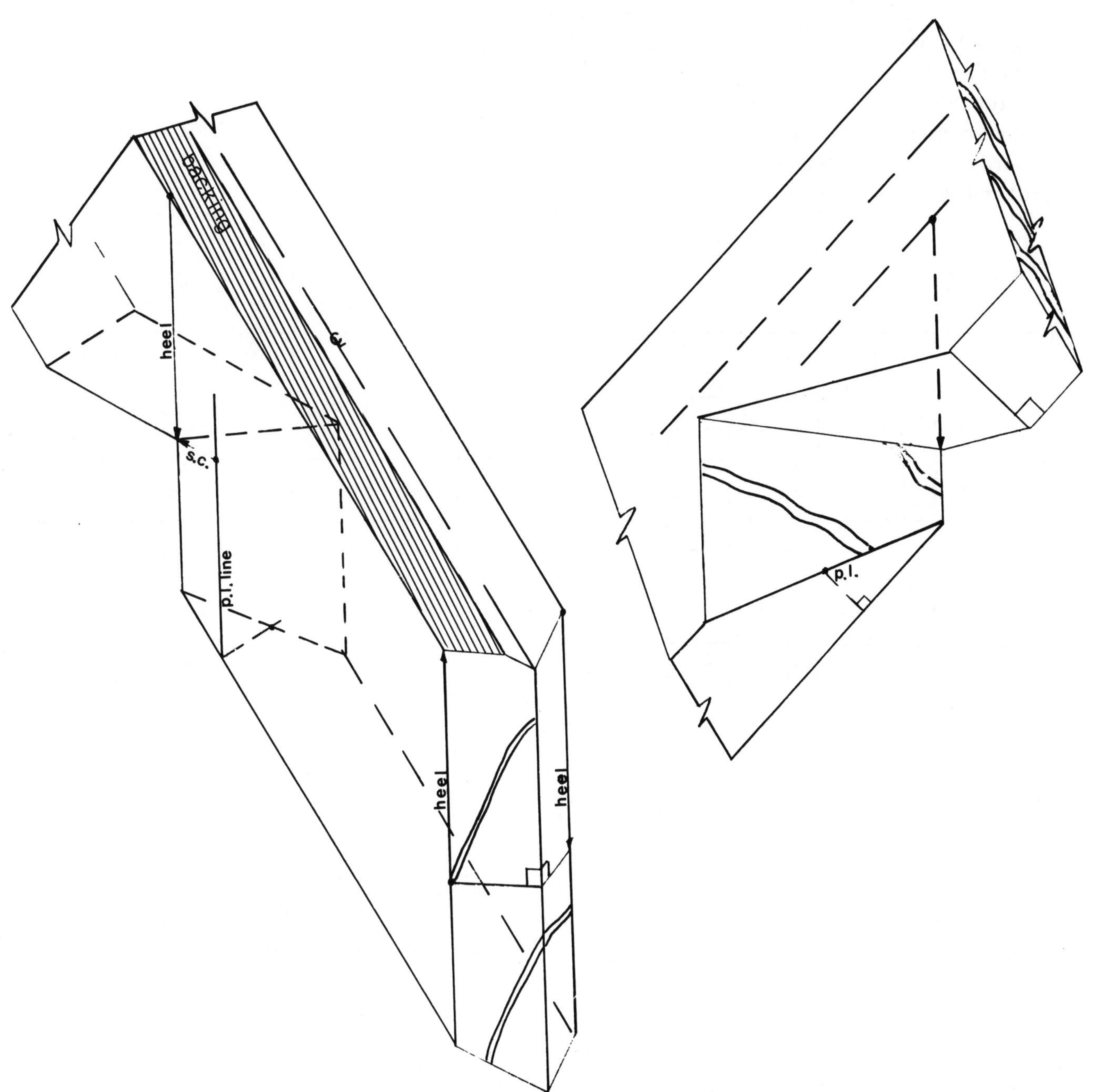

Truss Lay Out

7. To develop truss shapes for a simple roof, first layout the plan and two elevations. Then layout the desired centres. The common truss is marked out first with point length and bevels 1-2. The hip trusses are then projected from the plan, each one being like the common truss with the top off. Jack trusses are used at the lower hip end. These are the tail portion or the common, (see 0 shaded) and will be placed at 90° to the last hip truss. Another form of jack truss is planted on top of roof sheathing to form a roof extension. The profile of these is seen in the front view and the bevels identified as 3-4-5-6. The bottom chord must be bevelled to fit the main roof as at 7. Trusses should conform to "National Building Code."

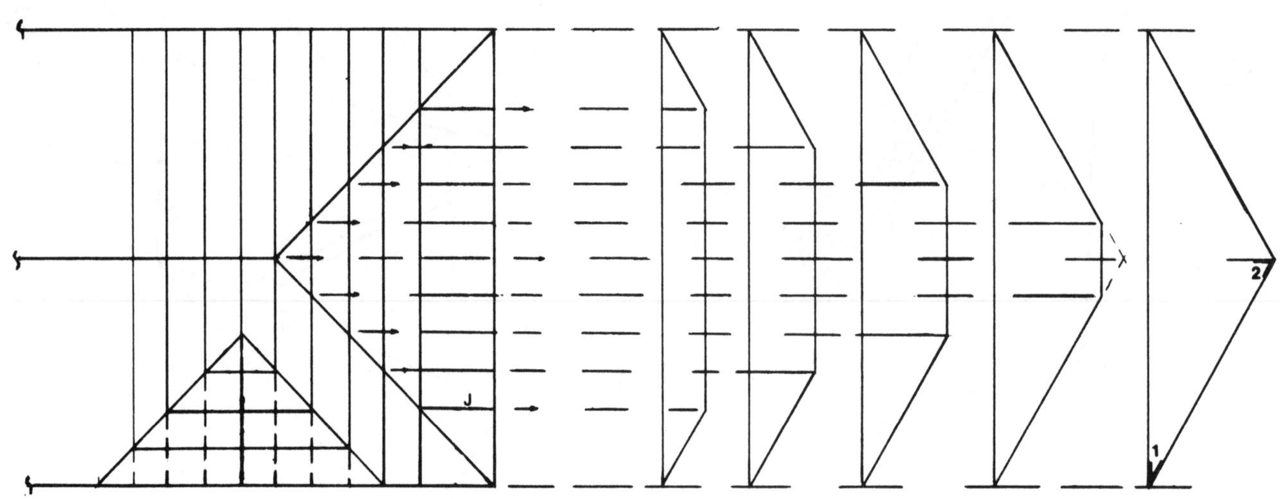

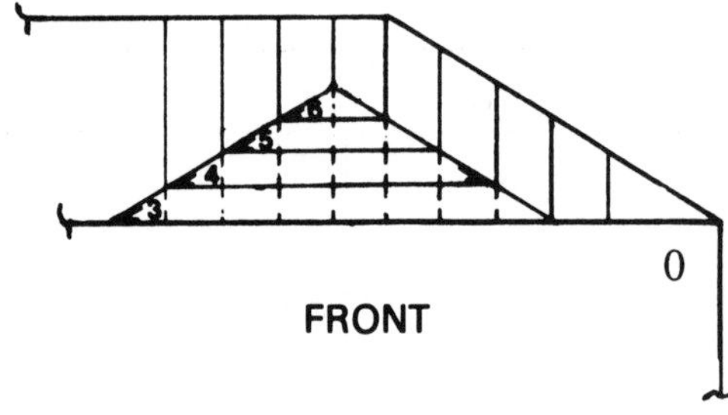

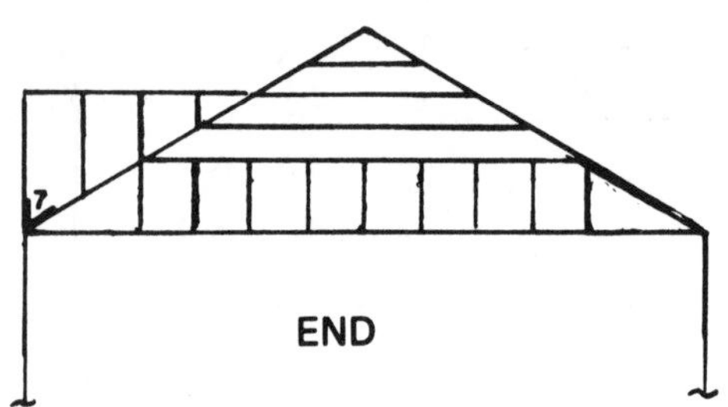

Bay Window

8. This drawing shows the development of a bay window roof:

First draw the plan and elevation. Next, use roof rise XY and lay out the common rafter on the plan showing its point length, plumb and level cuts at 1-2. Follow those steps on the plan of hip for 3-4. For jack rafter length and cuts 5-6, use its rise UV. Now develop surfaces E and F to give 7-8-9-10 hip side cuts and 11 jack side cuts. 13 and 14 are the dihedral angles (backing bevel) and 15 is the mitre angle for wall plates on the building face.

Scale 1:10

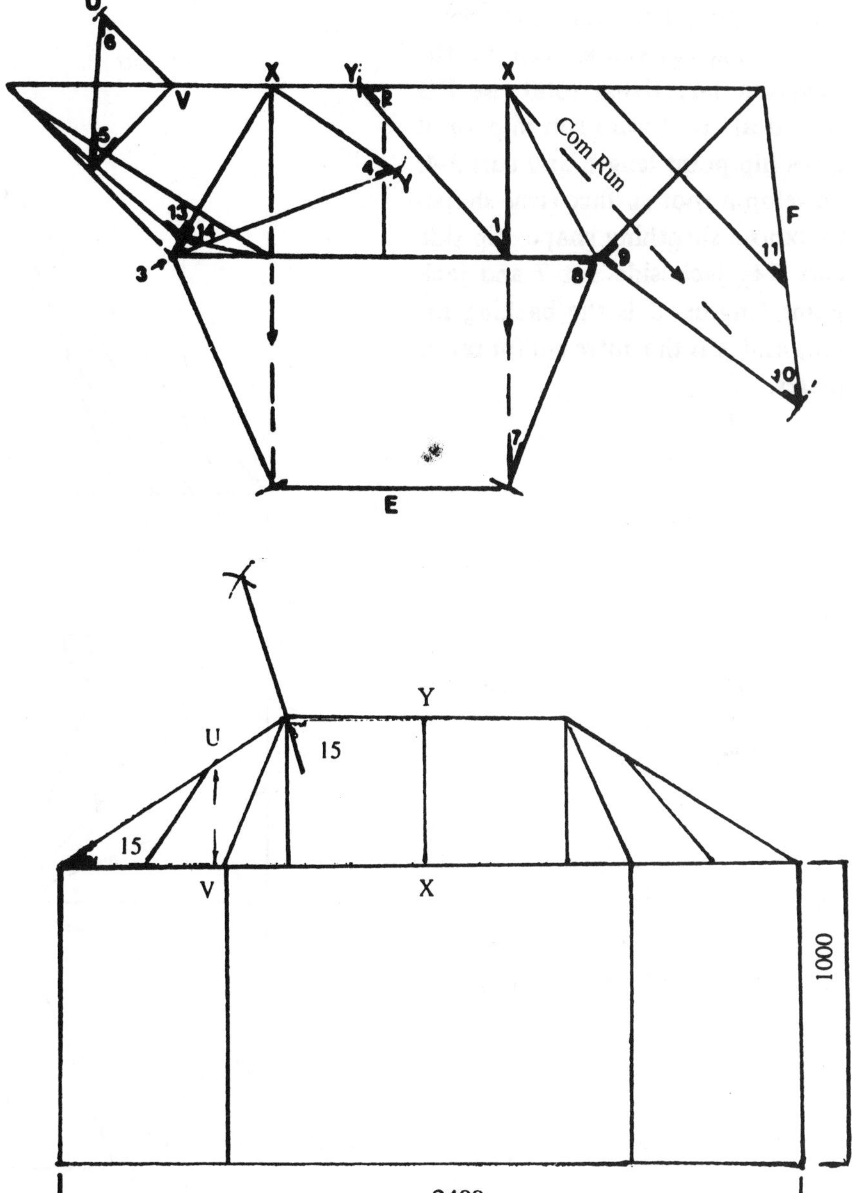

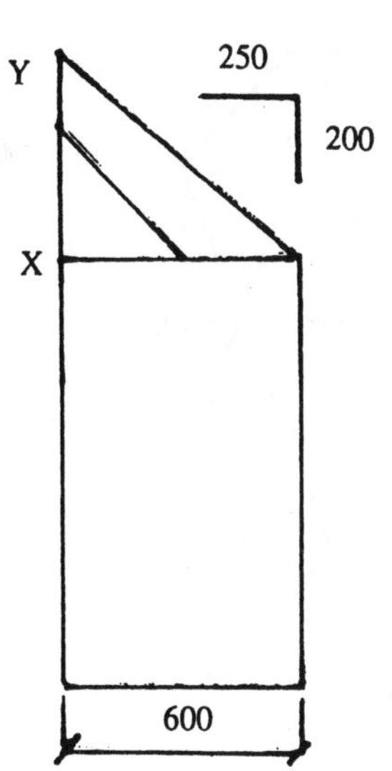

Ornamental or Tower Roofs

9. An octagonal roof has a common rafter run from side to centre QP. The common point length and cuts are seen in the elevation 1-2. Using total rise PO placed at 90° to hip run in plan, it gives hip point length and cuts 3-4. Develop a roof surface (true shape) to expose sheathing shape, hip side cuts 5-6, jack side cut 7 and jack point lengths. 8 is the backing for hips and A is the mitre cut for rafter plates.

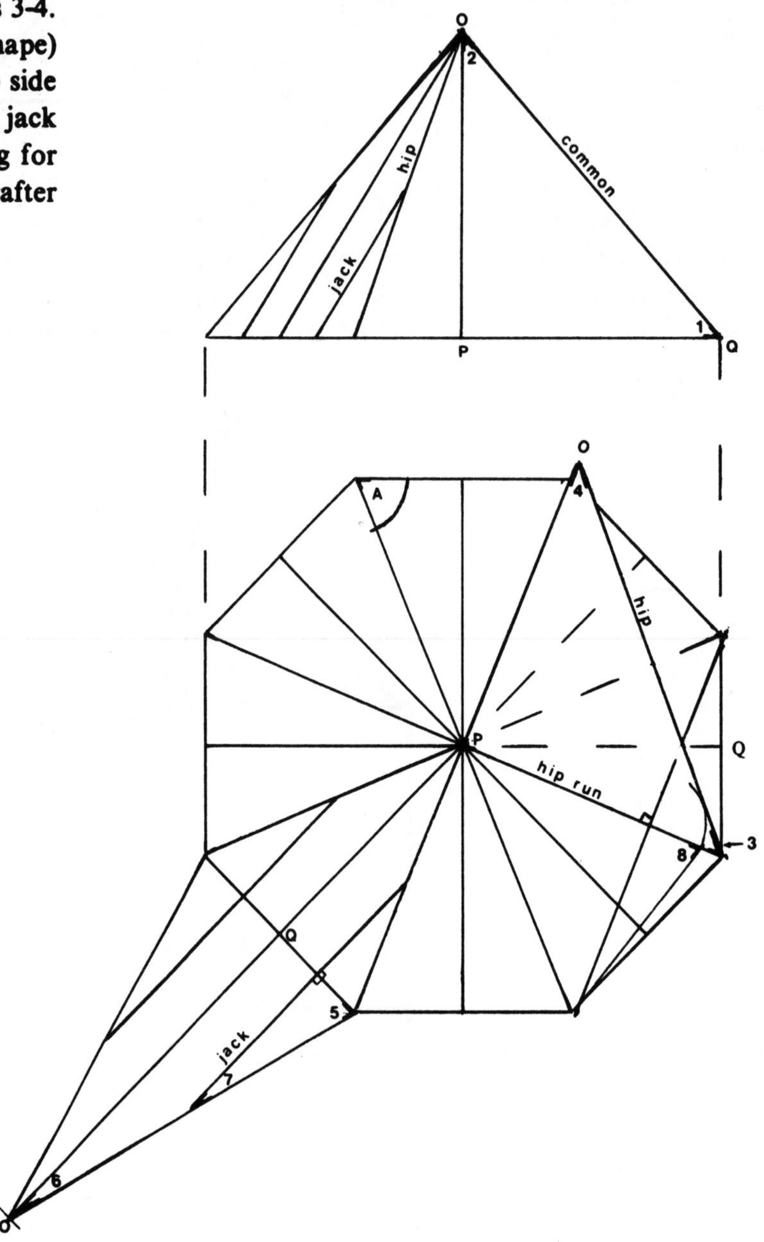

10. Ornamental roofs are many shapes, this one is an ogee on a hexagon plan:

The common rafter is the basic shape 0-6^1 and 0 to 6 the run. To develop the corresponding hip, project all points 0-6 from elevation to hip run on the plan. Then heights 1-1^1, 2-2^1, etc. are layed off at 90° at 1-1^{11}, 2-2^{11}, etc. Join 0-1^{11}-2^{11}-3^{11}-4^{11}-5^{11}-6^{11} to form the hip contours. Now develop the sheathing shape by using a compass to transfer points 0-1^1, 1^1-2^1, 2^1-3^1, 3^1-4^1, 4^1-5^1, 5^1-6^1 and lay them off at 90° from the plate (lower left). Project points out from the hip and connect the intersecting points.

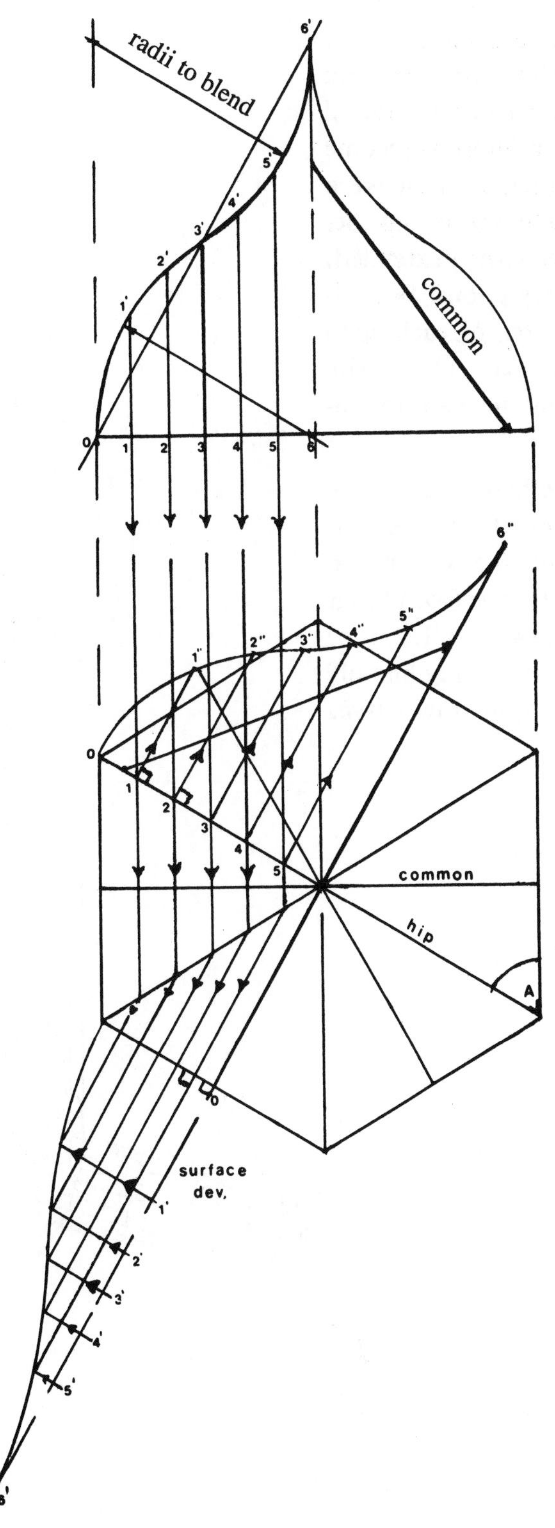

11. This diagram shows a dome or semi-sphere and can only be covered by close approximation. The surface would be developed as gores or by zones.

First layout a plan and elevation, then divide the curve into equal spaces as 1-5; these divisions become the zones. For each zone, strike a line through points ab, bc, cd, de. Let each line meet with 5y extended. From these intersections W,V,U, develop one quarter of each zone finding its length from the concentric sections in plan. (as, e-y[1])

For the gore method, project the intersecting circles at one section. Along the centre line 0-5[1] step off the distances over the curve 0-5. With a radius point at 5[1] swing 0, 1, 2, 3, 4, across to its corresponding line off the plan and layout the curve.

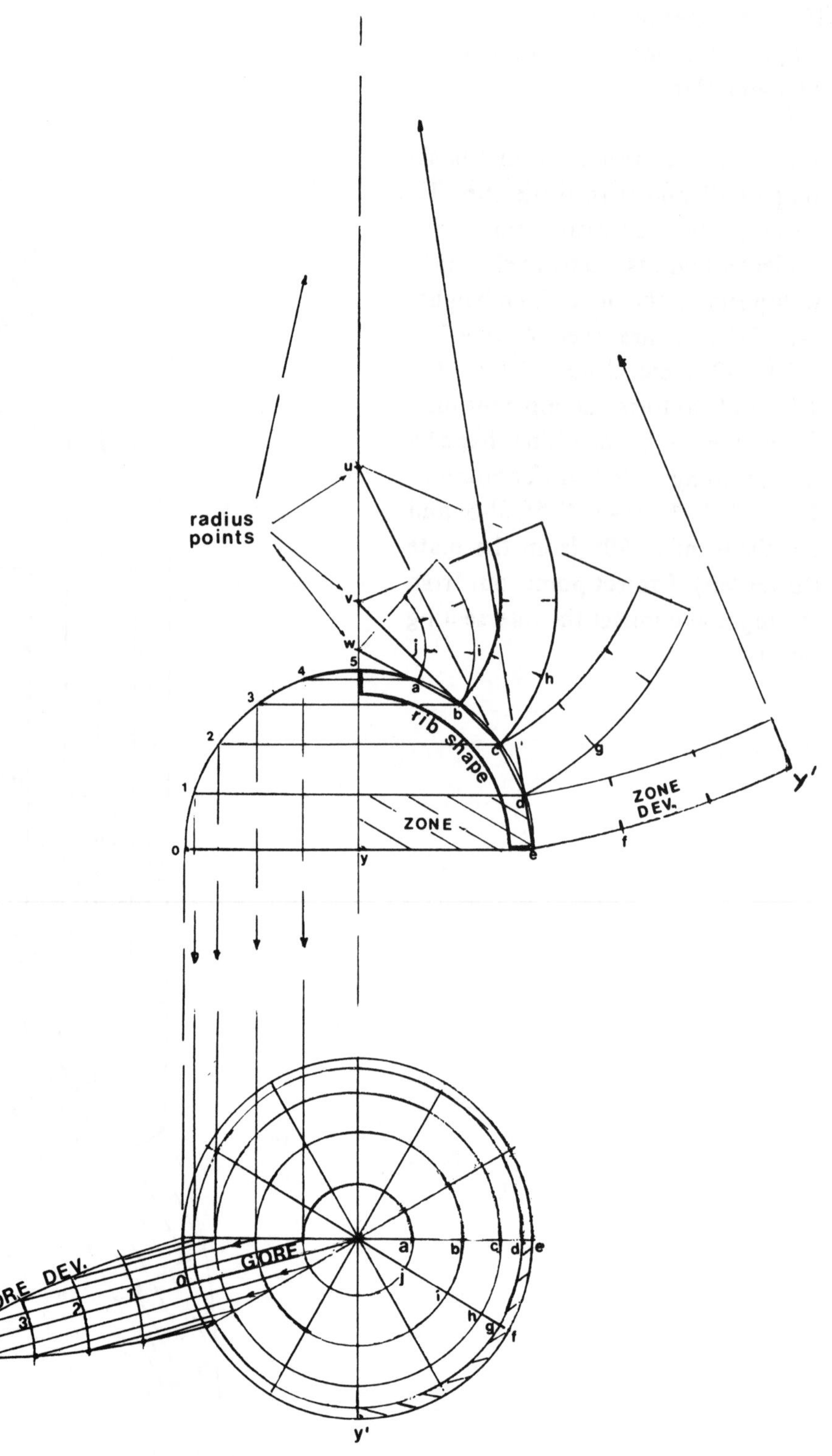

Related Roof Problems

12. A sawhorse is shown here because of the similarity to the roofing concept. By using ON from the end elevation at O[1]N in plan, the face bevel of the legs is exposed. By taking OM and laying off at O[1]M the edge bevel is seen. The leg length can be picked up from either development. 1 shows the bevel for housing into the header. 'A' depicts the method of cut at the housing. This type of sawhorse is sturdy when made with 2" x 4" material and properly braced.

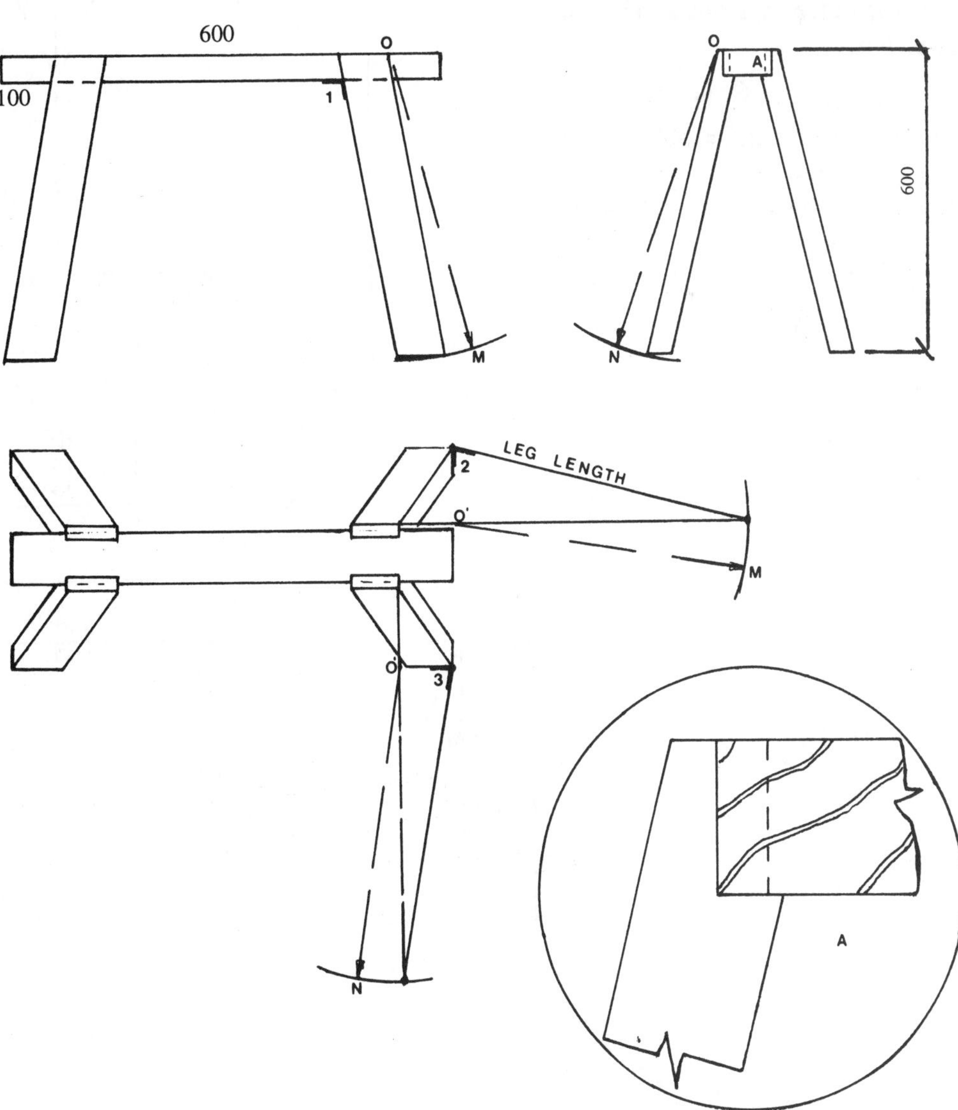

Related Roof Problems

13. A hopper is like an inverted roof. The method of finding the correct bevels as dealt with here should satisfy most conditions.

Develop one side to give 1 the face bevel (much like a sheathing cut). Then project ab at a′b′ and join b′ to the corner to give 2 the mitre angle. If a butt joint is desired, swing c to c′, step off half of ef at g. This gives 3 the butt angle. 4 shows the bevel for levelling the top edge which, if done first, the mitre lays out as 45°—see lower left.

Scale 1:5

2"=1'-0"

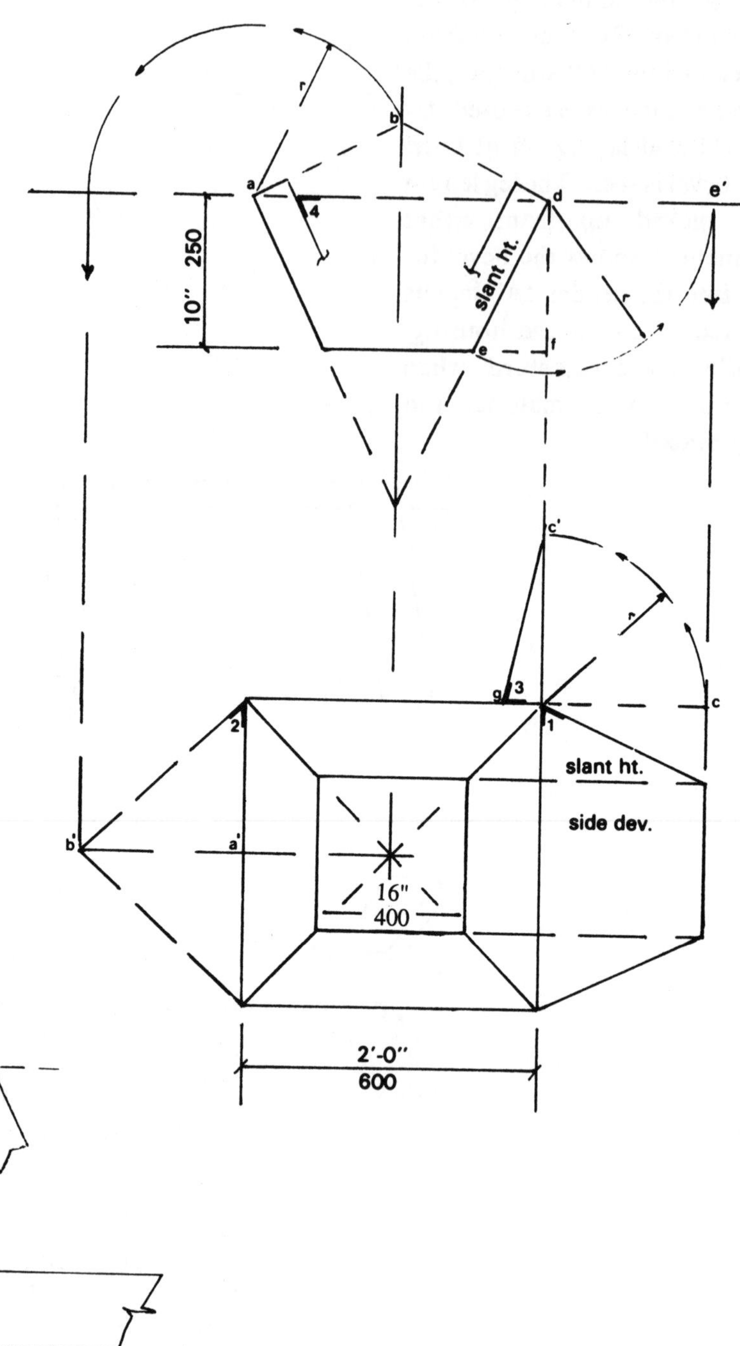

Related Roof Problems

14, 15. In these diagrams, only a part section and the mitre portion of hopper plan is used. T = thickness of material. 1 is the edge mitre. 2, the face bevel, is found by taking AB down to the plan, where line B is intersected by the plan mitre, and joined to C. For 3, the butt angle, let the material thickness T^1 project to d, then join d to e.

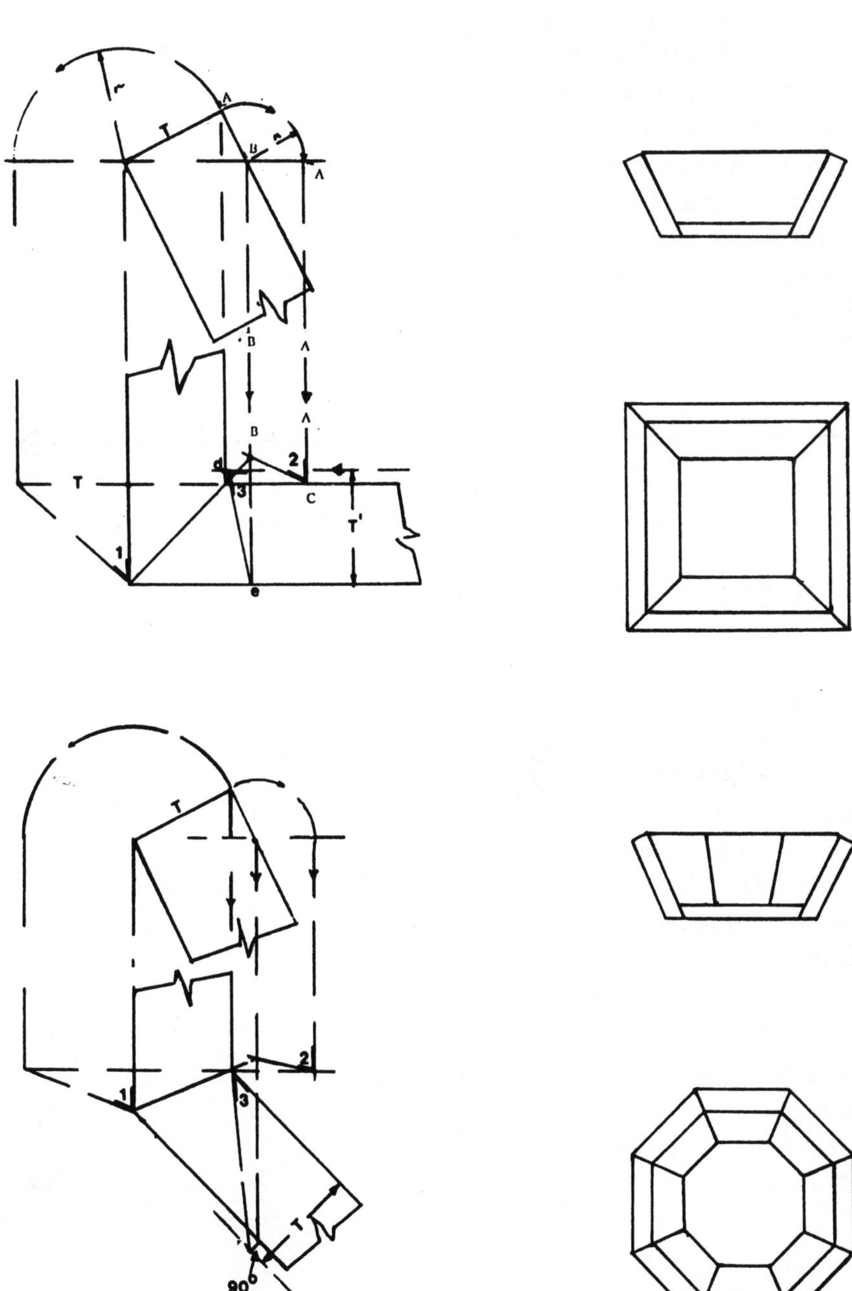

Related Roof Problems

16. Splayed window-door linings:

In this case the linings are splayed at 60°. Use 1 as centre a, to swing the lining face 1-2 into the elevation to meet a line produced from the head at 4. Joining 4-5 gives B the face bevel. The face edge bevel is seen at A thus making 90° over that edge for a joint as seen in the elevation. If a mitre is desired, strike a line anywhere across 5-6 at 90° to meet the edges at 0-0. Transfer 5-4 (joint length) to 6-7, cut by a 90° from 5. Now strike a tangent to make the dihedral — half of which is C the mitre.

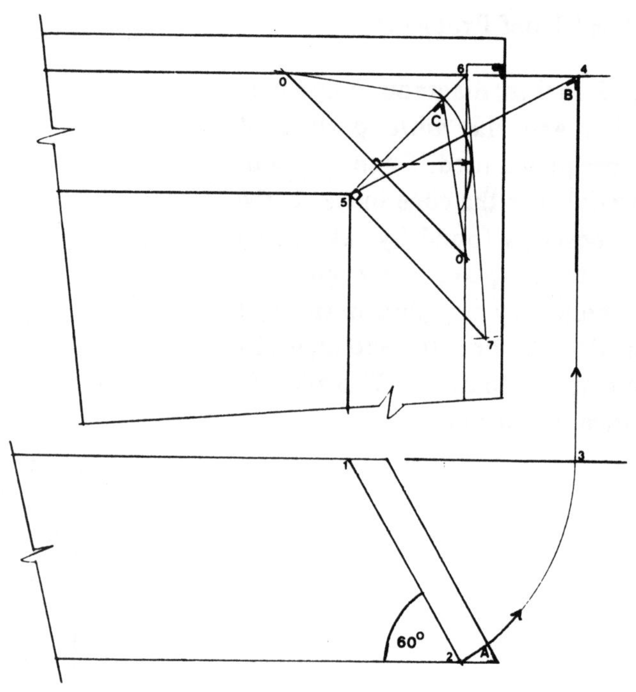

17. Triangular louvre board bevels:

Layout an elevation and section, then draw a vertical line through a. With a as centre, produce board width to b. Project b to intersect a vertical from d (depth of dadoe). Connect c to e for x, the face bevel. In a similar manner, produce board thickness to f, then to meet a vertical from the bottom edge of louvre at g. Connect e to g for y, the edge bevel. The dadoe layout for inner face of frame is shown on the frame development, z being the angle for dadoeing.

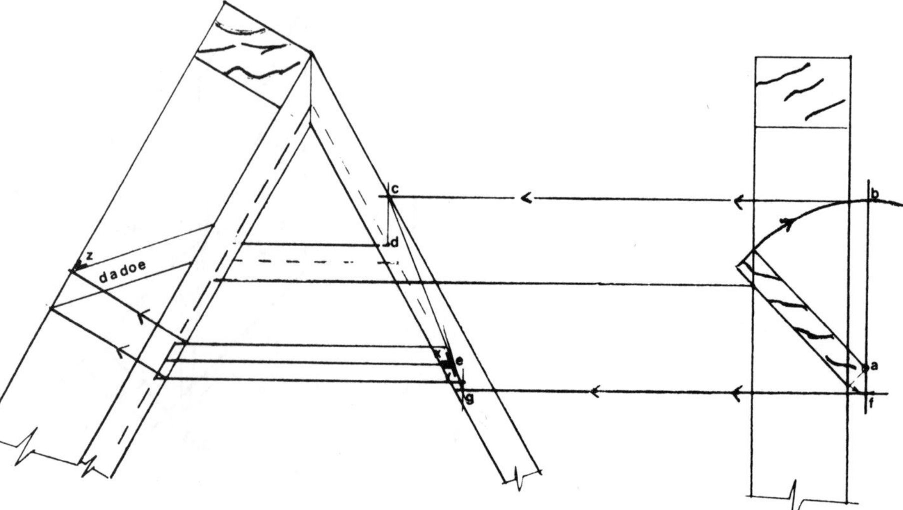

Discovery Exercises

1. To a scale of ⅜" to 1'-0" (1:50) redraw the plan and:

a. Develop surfaces A, B and gable C.
b. Show plumb and level cuts for common rafter D and hip rafter E.
c. Indicate where point length of D and E could be scaled.

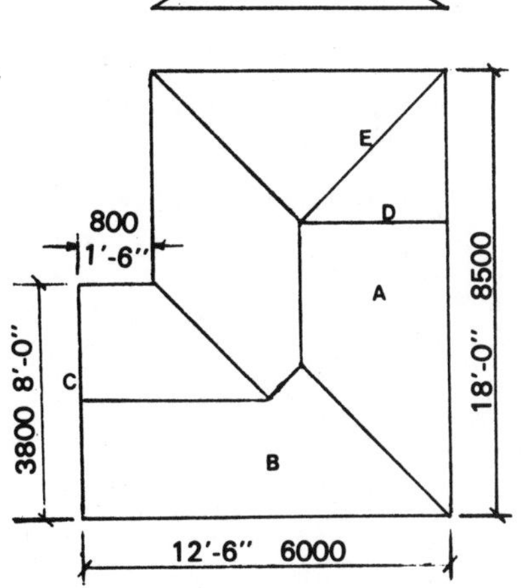

2. To a scale of ¼" to 1'-0" (1:50) redraw the plan and:

a. Develop all the rafter bevels as shown.
b. Write a key listing the name of each bevel.
c. Develop one surface and indicate on that development the point length of rafters a and b.

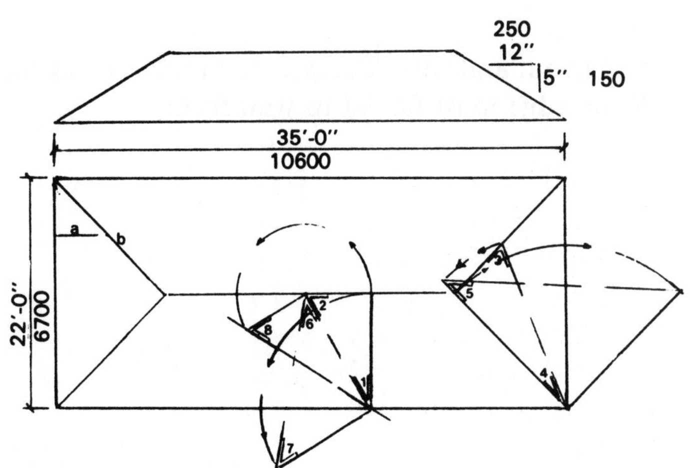

3. The shown plan has an unequal pitch roof. Develop the following:

a. Point lengths and cuts for rafters A,B,C,D.

b. Show by how much the rafter plate of the minor roof has to be raised to construct a level soffit with the same projection around the building.

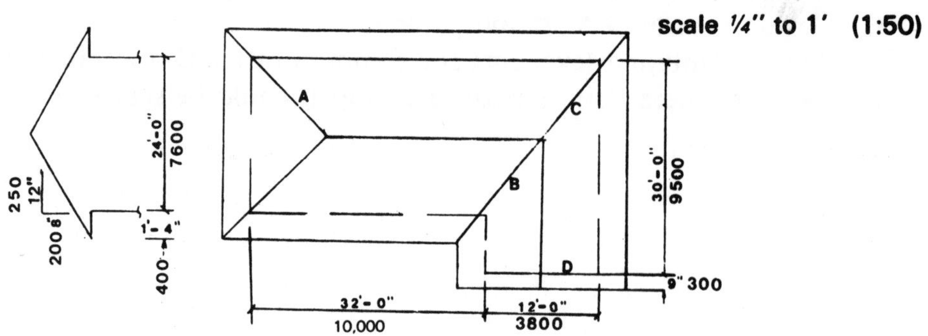

4. Develop the hip rib shape and one surface of the semi-circular dome.

scale ½" to 1' (1:20)

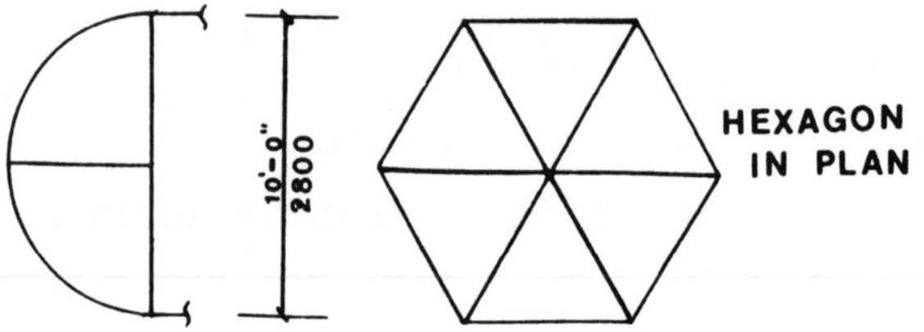

5. Layout and show the face and mitre bevels for a hopper, to form a recess in a concrete wall; ½ size. Note: edge must fit flat to wall form.

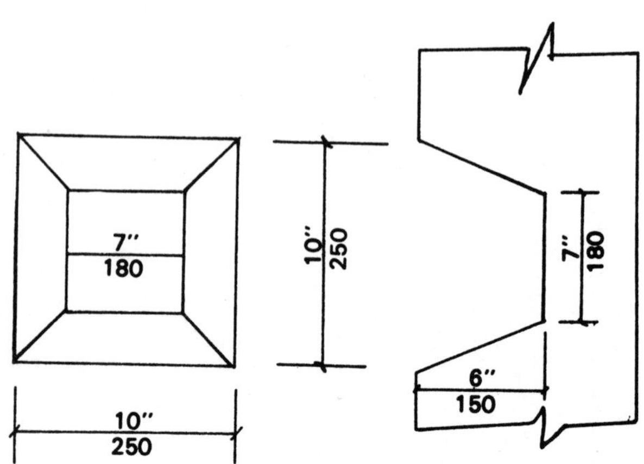

Roof Problems

Unit 11
Surface Shapes and Auxiliary Views

The objective here is to layout the true shape of a surface or surfaces, so they may be cut from a material to construct or cover a given shape.

1. Layout the plan and front elevation of the given prism. Starting with vertical corner cc[1], visualize the box turning to the right along line cc opening the flaps like an envelope. Layout each surface following prism dimensions and project lines accordingly. It is important to letter as you draw, so as not to lose track.

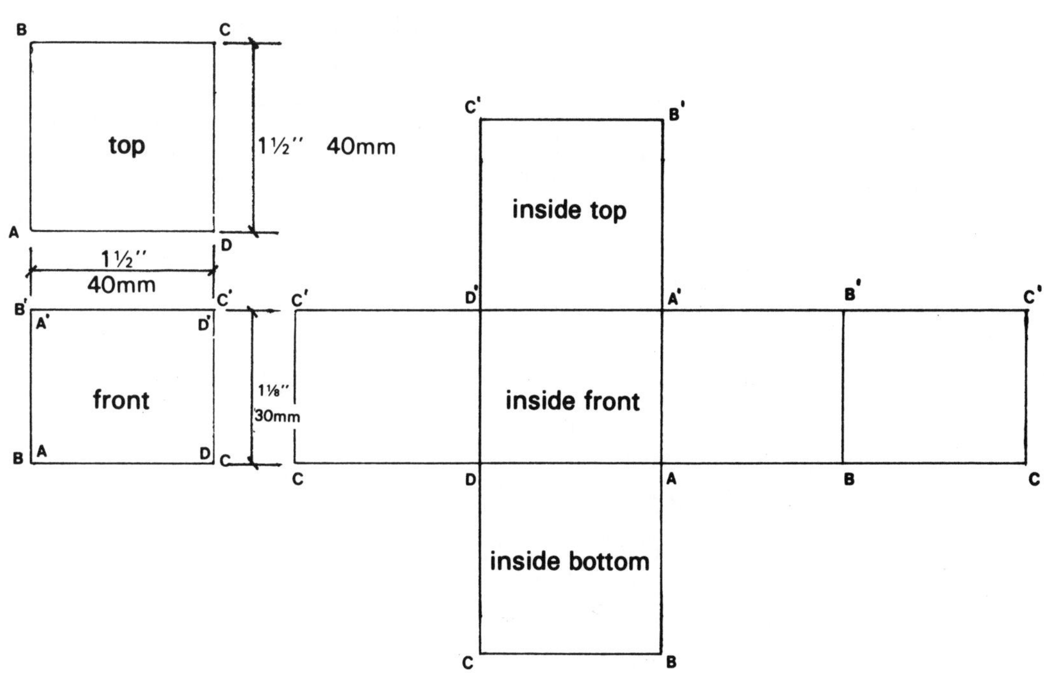

2. Layout the plan and elevation of the cylinder. Divide the plan into sectors with a 30°-60° set square (although the more divisions — the greater degree of accuracy). Draw a stretch out line to the right of the elevation. With a compass, pick up the distance between sectors on the plan 1-2, 2-3, etc. Step off the same number of these divisions along the stretch out line, thus measuring the circumference. Lay off the vertical height and complete. You have unrolled the cylinder and now view it as one flat surface.

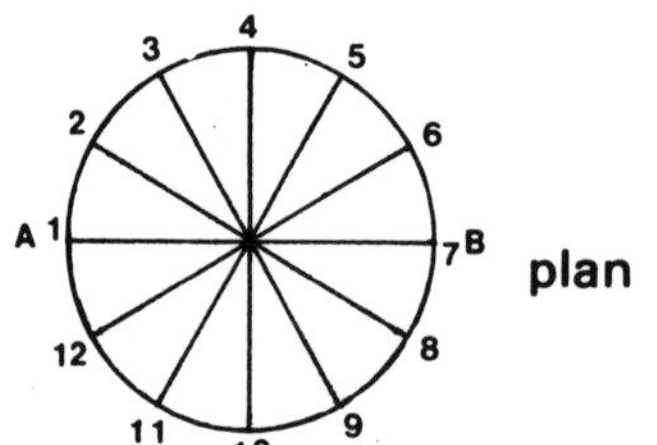

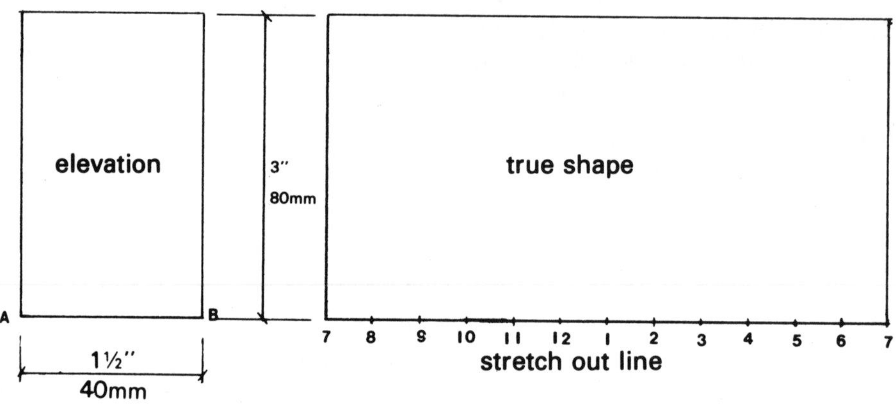

3. To find the shape of an intersecting plane and the surface development of a prism:

Draw the plan and front elevation. Then at right angles to the cut, project AB and CD. Between these lines, construct the true shape of cutting plane, known as an auxiliary view. A^1D^1 is the same length as AD in front elevation and A^1B^1 is the same as AB in plan.

Next, to the right of the elevation, draw stretch out line 3-3. Set out faces 3-4, 4-1, 1-2, 2-3, taken from plan. Erect light perpendiculars at these points. Project the corner heights across from the elevation to corresponding lines. Join the points $C^1D^1A^1B^1C^{11}$ to complete the development.

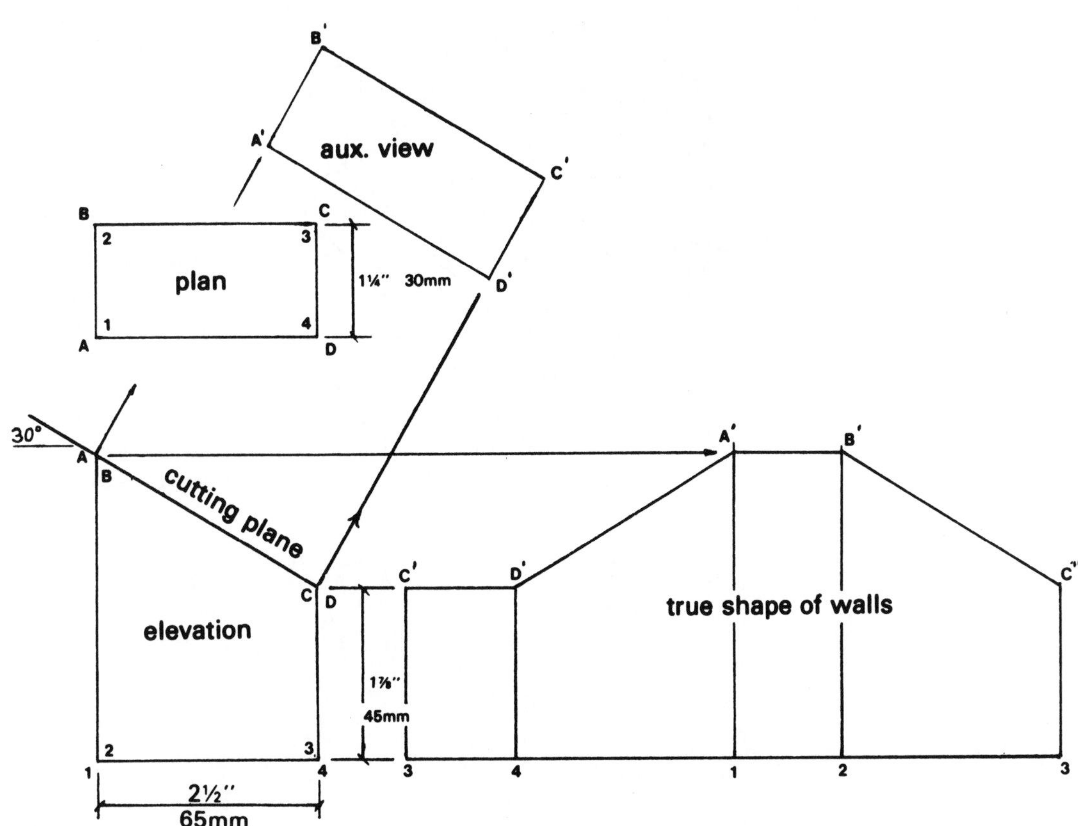

4. To find the shape of an intersecting plane and the surface development of a cylinder:

Draw the plan and elevation. Divide the plan into equal sectors and number them. Project these points down to the elevation. The intersection is an ellipse whose shape is found by projecting at right angle to the cutting plane, light lines from the projected points 1-2-3, etc. The line xy in the elevation is the major axis, and the diameter of the cylinder is the minor axis. From the plan, transfer the measurements from line 1-7 to 2, 1-7 to 3, etc. over to x^1y^1 major axis and form the ellipse.

Next, draw a stretch out line and step off the numbered sectors. Perpendicular at these points, erect light lines, project across the corresponding heights, 1-2-3, etc. Connect the points to make a curved line which completes the cylinder wall development.

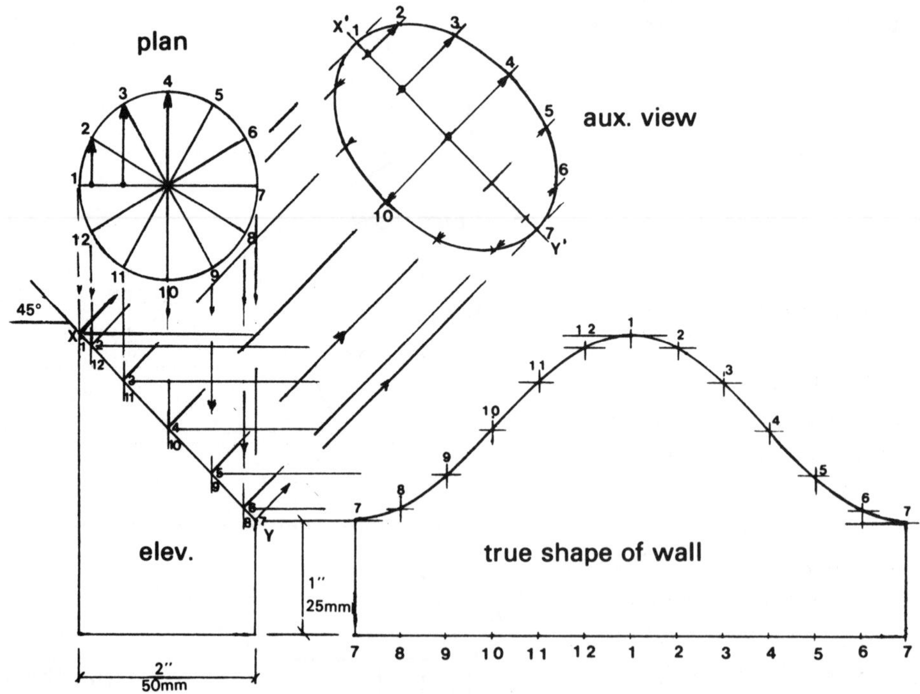

5. A chimney intersecting the ridge of a roof:

a. What shape and size hole must be framed?

b. What shape would a vertical continuous flashing be?

Start with plan and elevation and project an auxiliary view from one of the equal slopes. Further develop the wall surfaces to find the true shape of flashing.

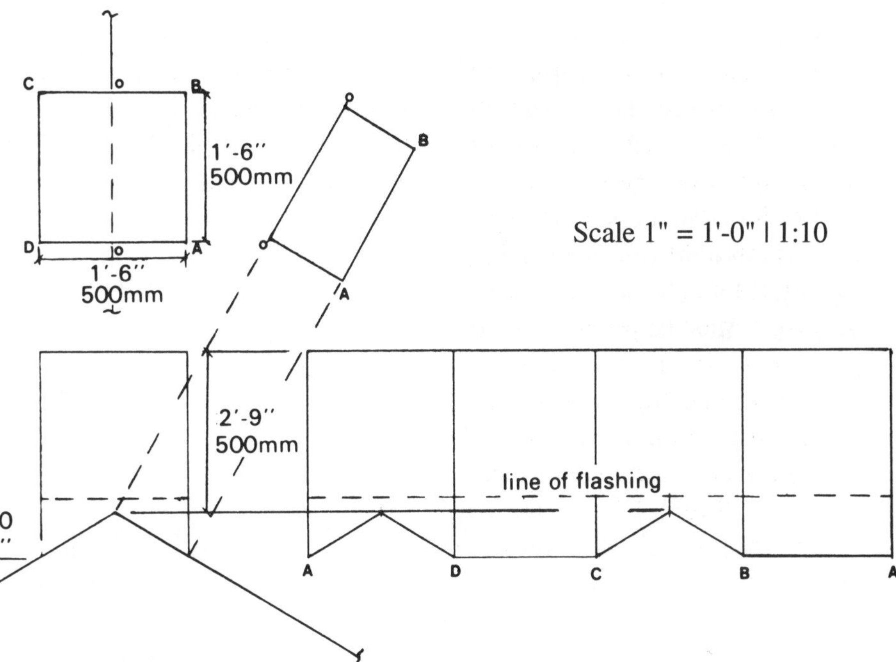

6. An intersection of ducting:

The true shape of pipe 1 must be developed to enable a fit of same at wall xy and joint ag[1]. Proceed as in problem number 4, but note the development is unrolled directly from the plan view. This is incidental. An auxiliary view would be required at xy so that an elliptical hole may be cut in the wall.

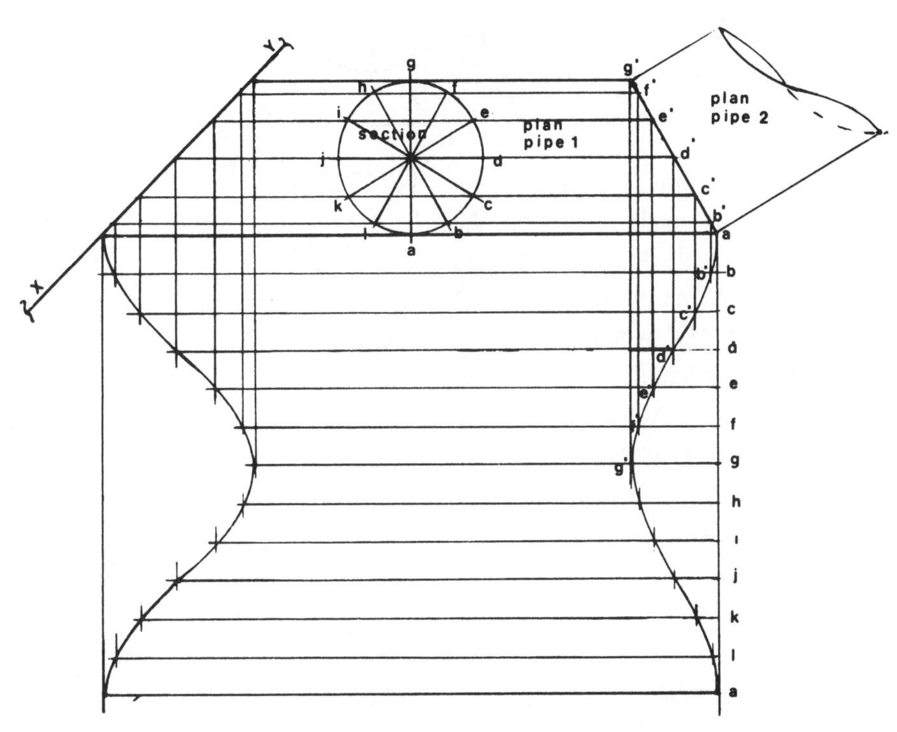

7. To find the true surface shape of a triangular pyramid:

The pyramid has a base and three sloping sides. The dimension of the base must be known and the vertical height as given here 00^1, or the slant height which would be O^1D. First, lay off 00^1 at right angles to BO on the plan. BO^1 is the true length of the corner (or hip). Use a compass to mark length BO^1 from A and C at 0^{11}.

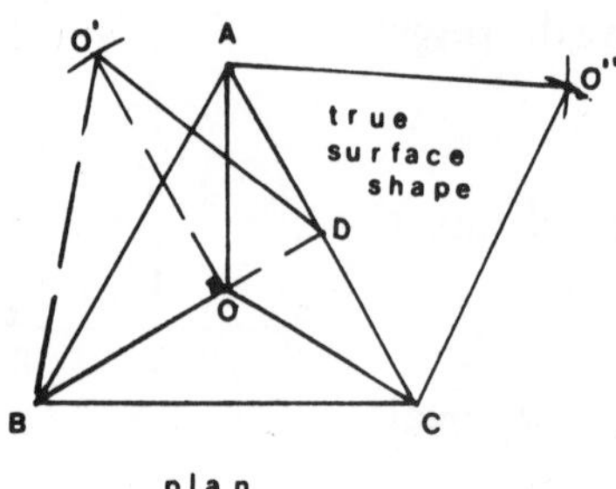

plan

AB=2½" | 65mm
00^1 = 2" | 50mm

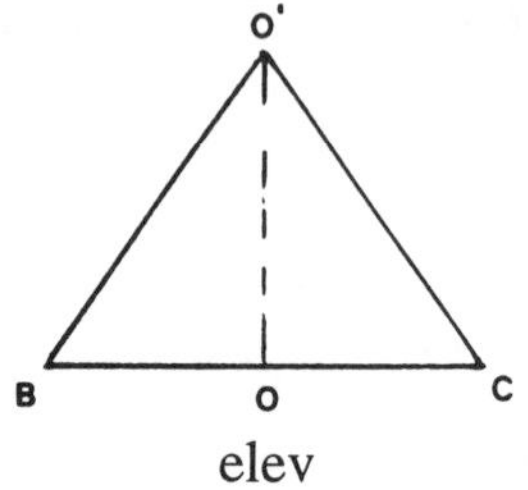

elev

8. Other pyramids may be developed in the same manner as in number 7:

With this hexagonal pyramid, the slant height is readily seen on the elevation as A^10^1; this was not so in number 7. Think in terms of roof framing, then AO^1 would represent a common rafter length, and CO^{11} the length of hip. FAO^{111} would then represent the true shape of sheathing.

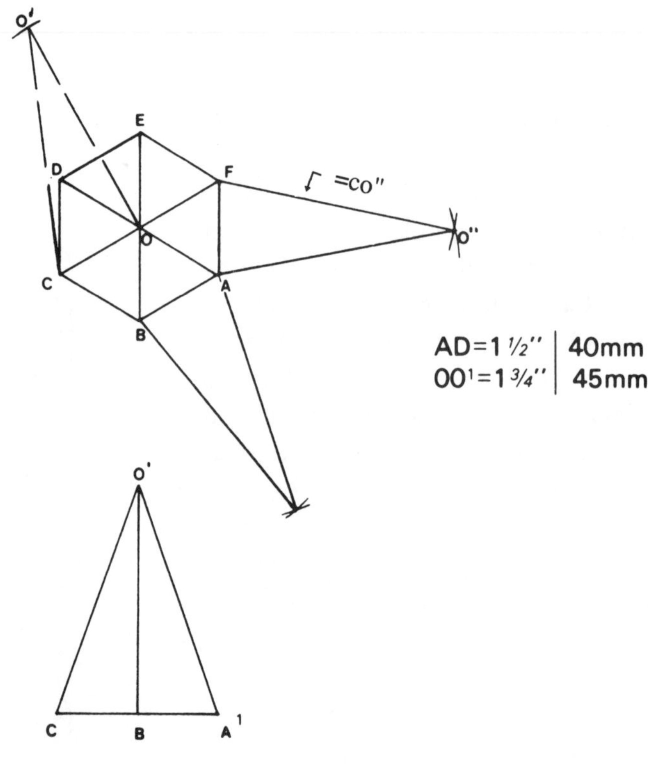

9. Development of a truncation and auxiliary view:

Developing the surface of a truncated pyramid in one surface is used when cutting all sides from one piece of material. This square pyramid is shown in plan and elevation with its top cut off at an angle. The length of the hipped corners must first be found. This is done by transferring OD to the elevation for $0'$-ABCD. Further, project abcd to $a'b'c'd'$ on true corner length $0'$-ABCD, this will identify where the cutting plane passes through each corner. Now using $0'$-ABCD as a radius, scribe an arc, step off pyramid base DA, etc. Pick up from the elevation lengths Dd', Aa', Bb', and Cc', plot them at $d'd$, etc., and join all points. The cut must now be shown in the plan, project points abcd up to meet OA, OB, OC, OD. Proceed to develop the true shape as an auxiliary view, $a''b''c''d''$.

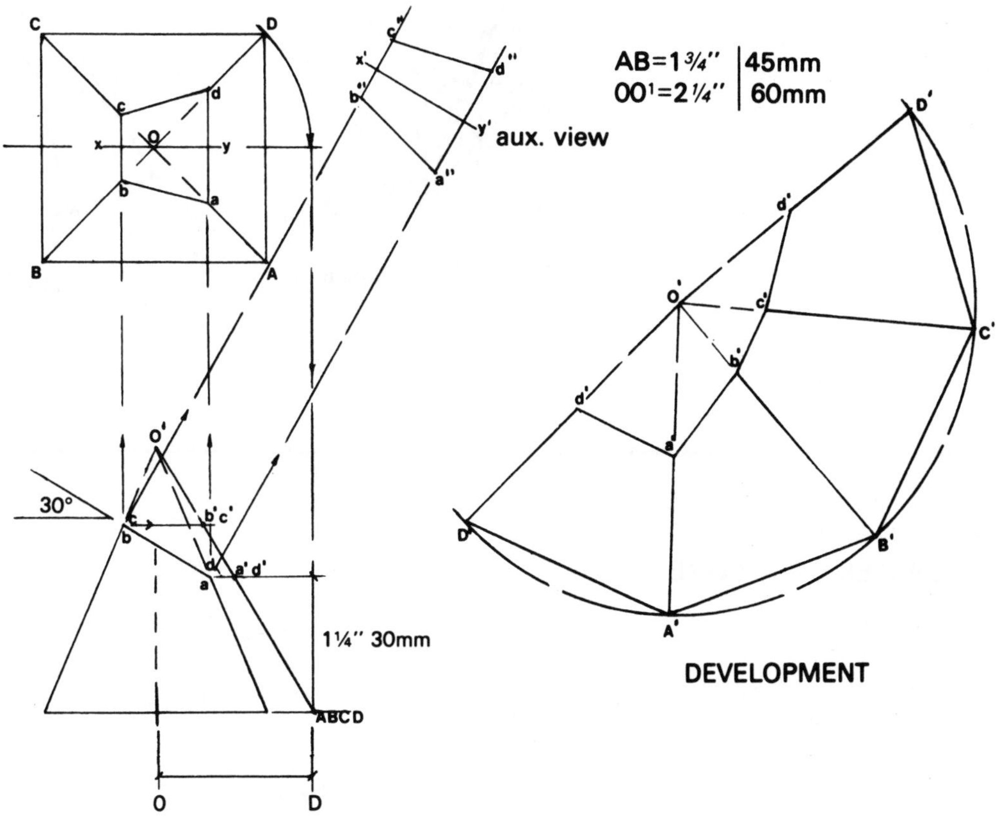

10. Follow the steps in number 9 to expose the true shapes of this octagonal pyramid. Remember the true hip length must be used to determine heights where the cut passes through each corner as at Aa^1 on the elevation.

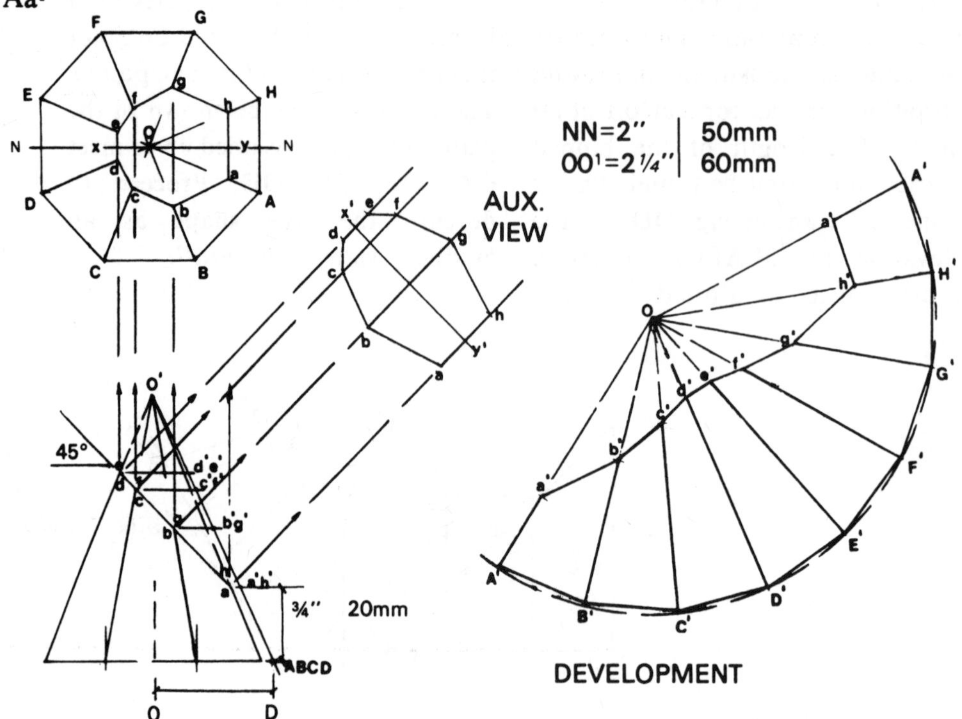

PROPERTIES OF A CONE

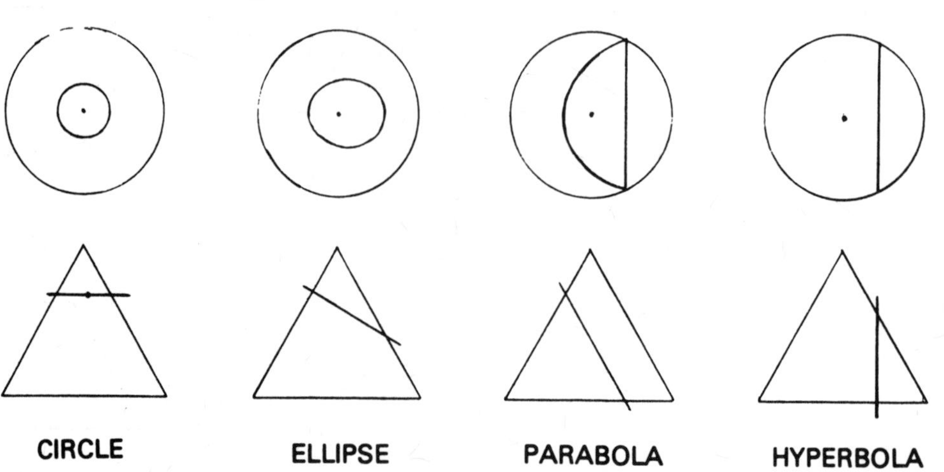

11. There is only one way to develop the surface of a cone; and that is on the same plane. The sectors from the plan 1-2, 2-3, etc. are stepped off on an arc to determine the length of circumference, the more sectors, the truer the length. The arc's radius is the slant height of the cone 0^1-1.

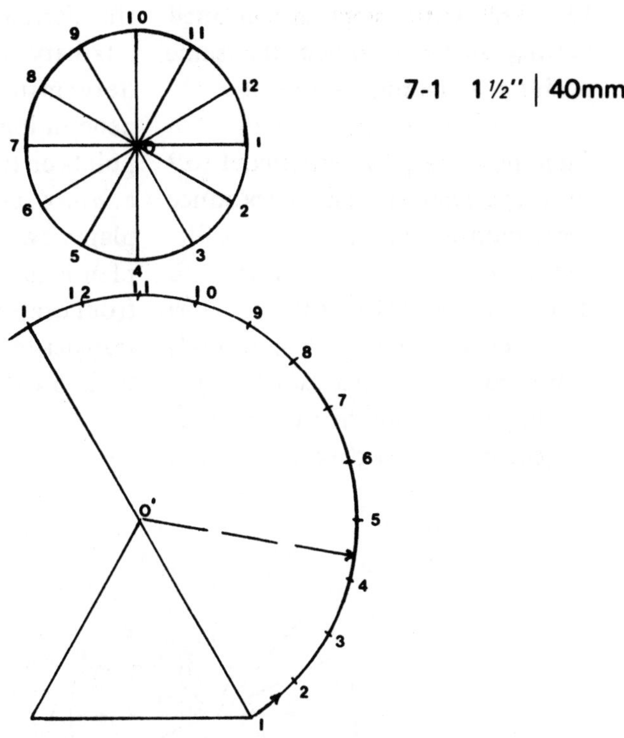

7-1 1 ½" | 40mm

12. Follow the steps in number 11 with an additional arc (radius 0^1X) to accommodate the slice off the top.

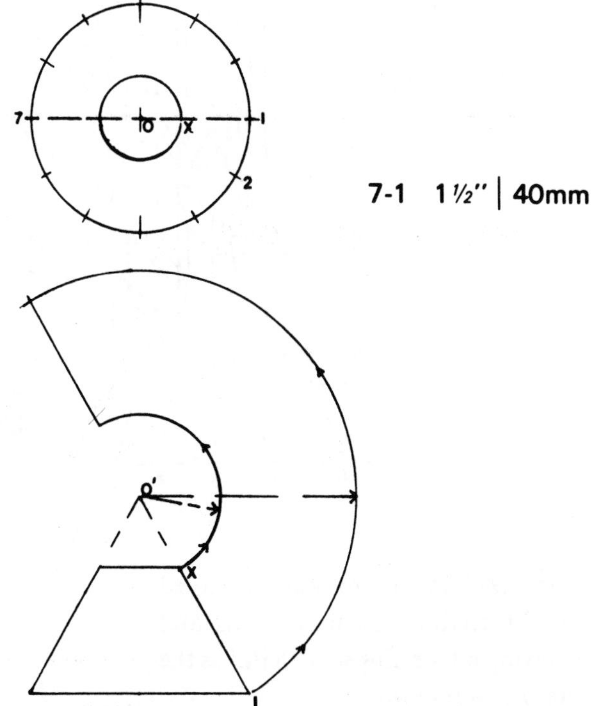

7-1 1 ½" | 40mm

13. The cone with an inclined cutting plane is much the same problem as number 9.

First, project the sector points 1 to 9 to the elevation base. Reproduce these points from base to apex 0^1; these lines are necessary to determine heights to the cut. The slant height is seen at 0^l-9 or 0^l-1, therefore the intersection of the sloping lines and the cut must be projected to abcdefghi, further, transferred to the development. The auxiliary view may be needed to determine the section of a penetrating tube. In which case points on the cut must be taken up to a, b, c, d, etc., on the plan to plot the plan view of the cut. This shape in plan is necessary to pick up widths from centre line to b and c, etc., for transfer to plot points on the ellipse as at $b^lc^ld^l$, etc.

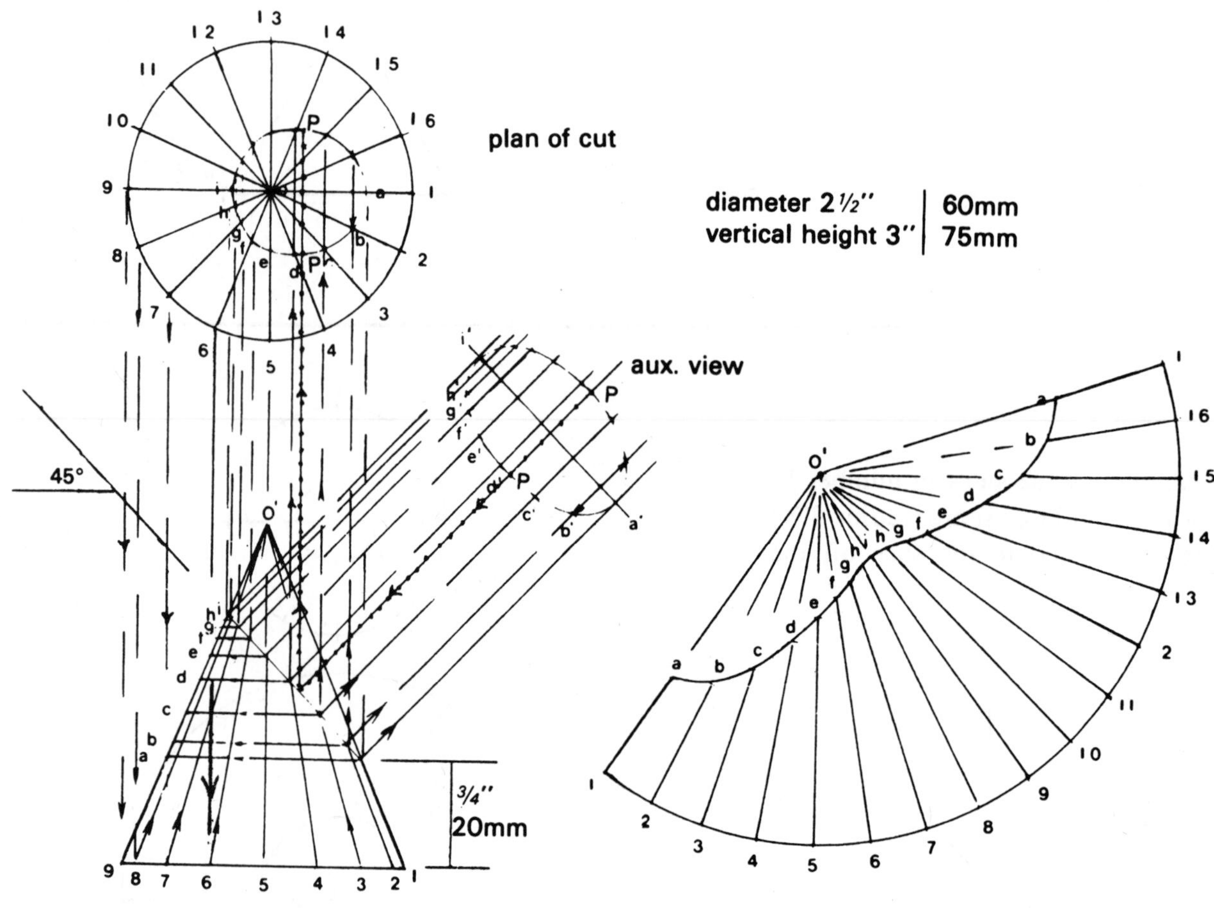

SURFACE DEVELOPMENT OF TRUNCATED CONE

The auxiliary view may also be layed out by finding the minor axis and drawing a true ellipse with i^la^l as the major. p-p=minor.

Numbers 14-17: advanced study.

14. This problem is to form concrete window suntraps. The wall is actually an inclined cut (R-P) through the lower portion of a cone, O-P being the cone base. Note only half of all developments are necessary. The auxillary view is used to layout the hole shape in the wall. The inner skin for this conical window would be found from a similar cone reduced by the concrete thickness.

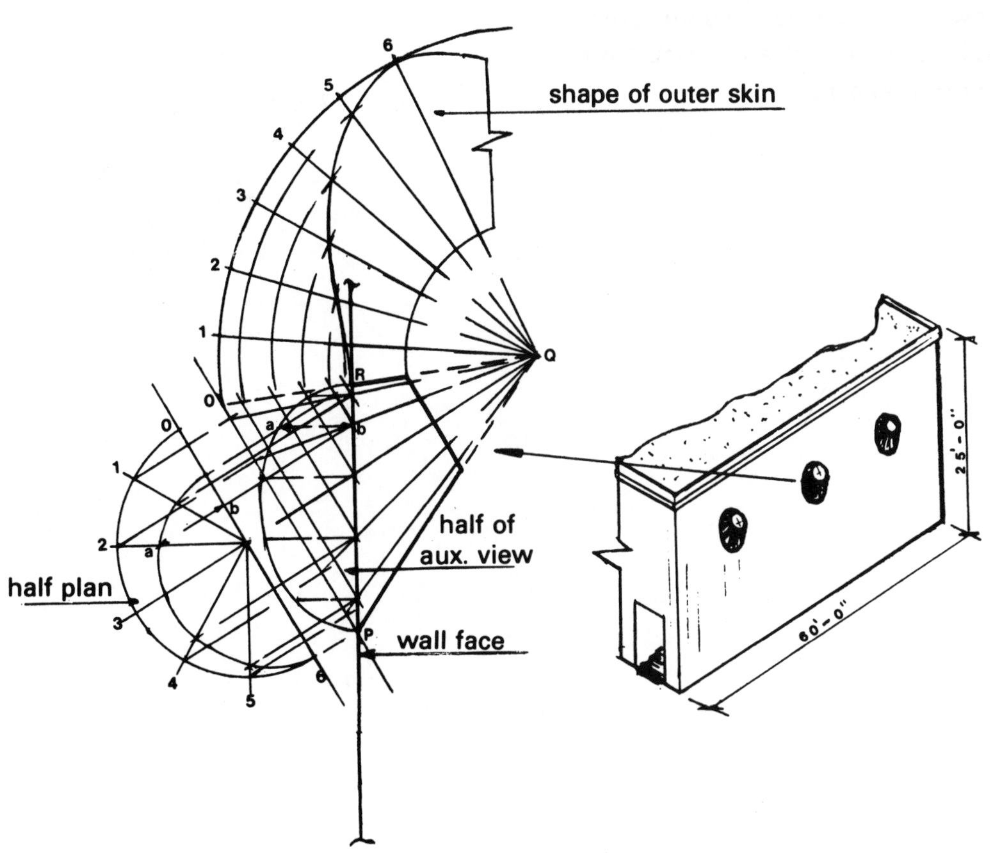

15. A semicircular splayed lining is a section of half of a right cone. If cone development is understood, then this layout should be readily seen:

First, draw the elevation and horizontal section of jamb linings. The cone which generates from C is reproduced from C[1]; using C[1]6 and C[1]D as radii for inner-face of lining, only a half development is necessary. If a pliable material is being used for linings, the face mould O and P may not be needed; however, if cut from solid wood they will be. O and P are struck from centre E and F.

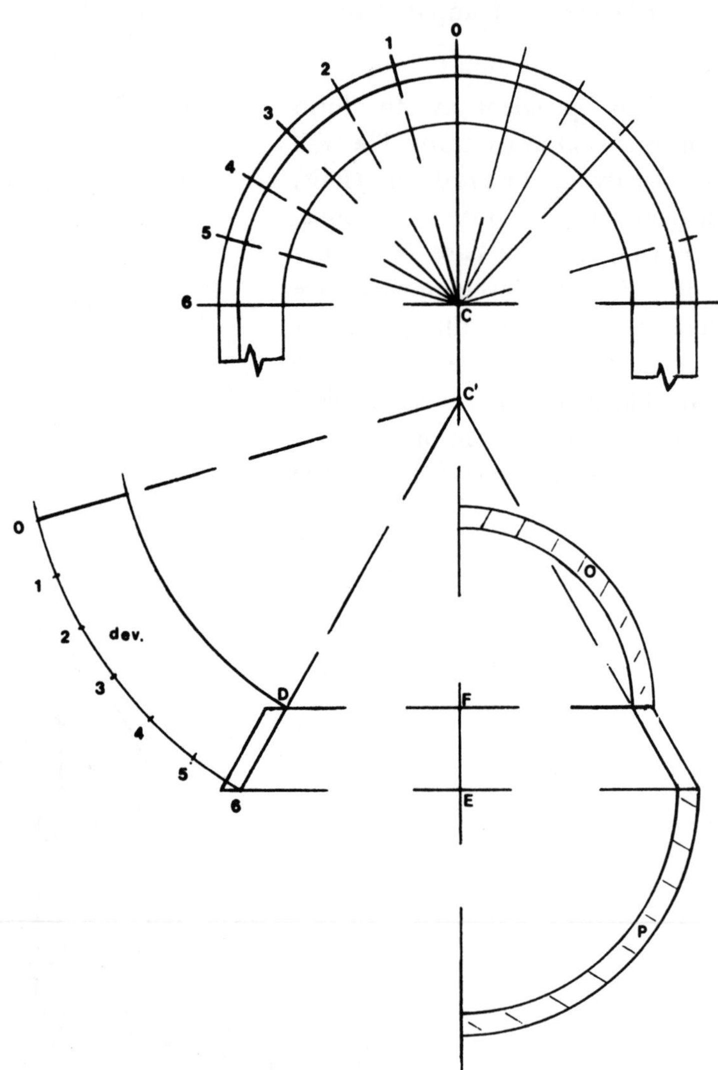

16.

CIRCLE ON CIRCLE
splayed jambs and splayed crown

17.

EYEBROW DORMER

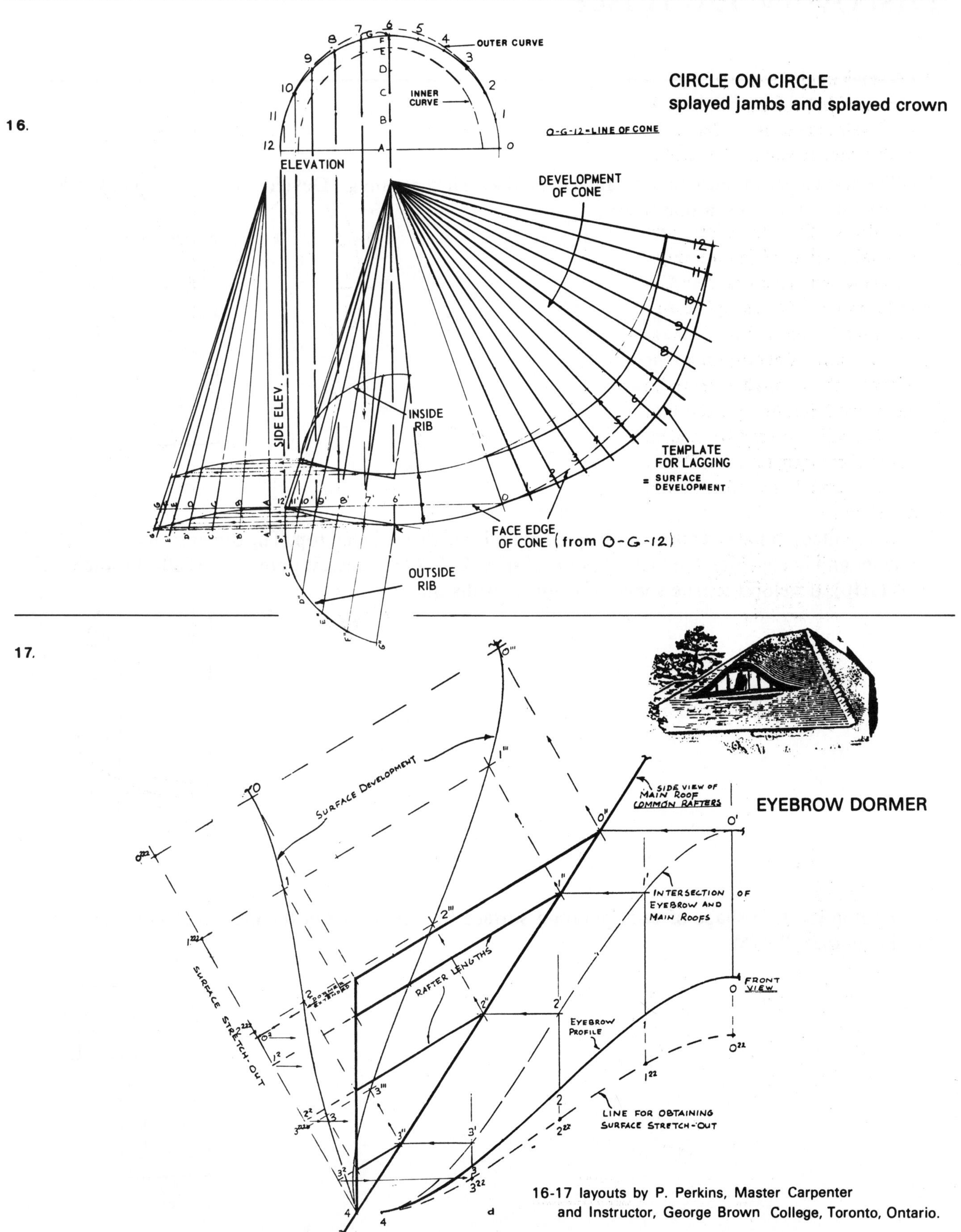

16-17 layouts by P. Perkins, Master Carpenter and Instructor, George Brown College, Toronto, Ontario.

Discovery Exercises

1. The shown prism must be built and covered with heavy canvas. Develop the shape of canvas required to cover the whole job in one piece.

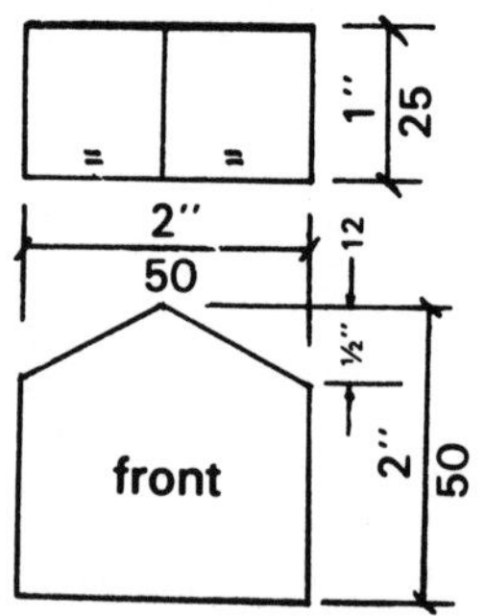

2. An architect's model requires a plywood cylinder with a splayed top. The dimensions to be 1'-0" (380) diameter and in elevation the highest point of splay 2'-0" (760), lowest point 1'-6" (560). To a scale of 1½" to 1' (1:10), develop the true shape of cylinder walls only.

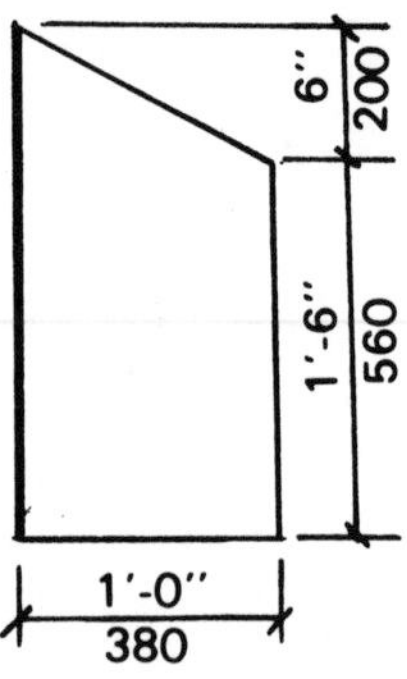

3. Develop the true shape of each pyramid surface by hinging from the plan as in number 8. Vertical height to apex 3" (76).

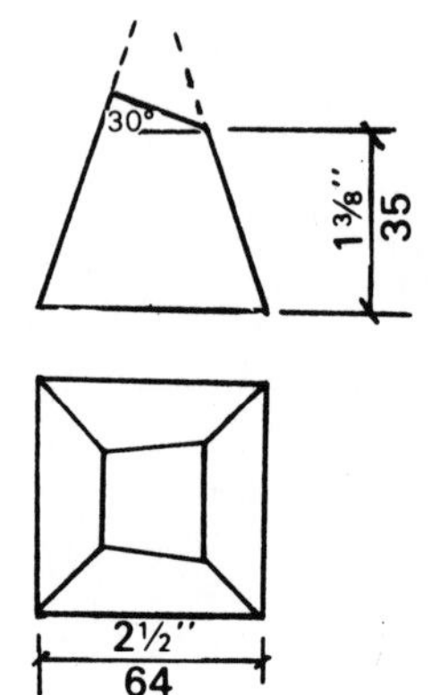

4. Find the slant height xy of the pyramid.

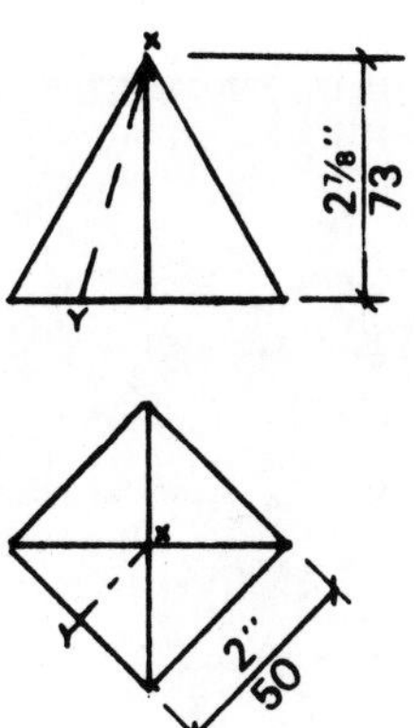

5. Develop the true shape of pyramid walls, hinging them from the plan. Also, layout the shape of the inclined cut. Scale 1" to 1' (1:20). NOTE: Care should be taken in finding true length of corners and the width of auxiliary section.

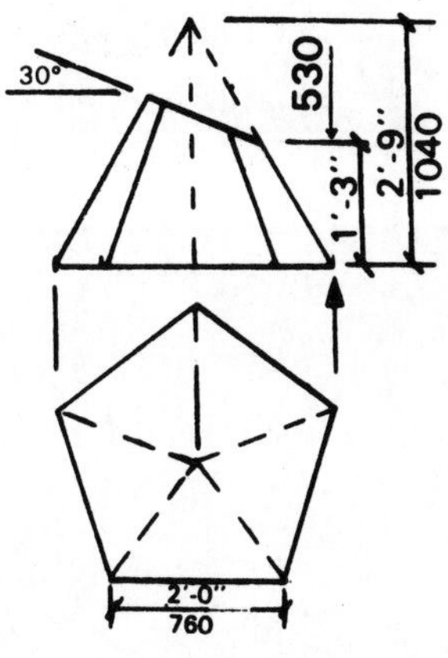

6. A stage prop has to be built with ribs and a plywood skin. Show the sectional shape of upper tube and develop the shape of cone surface. Scale ½" to 1' (1:20).

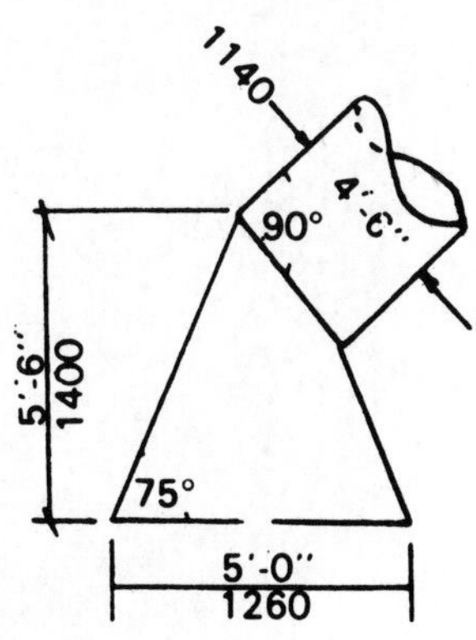

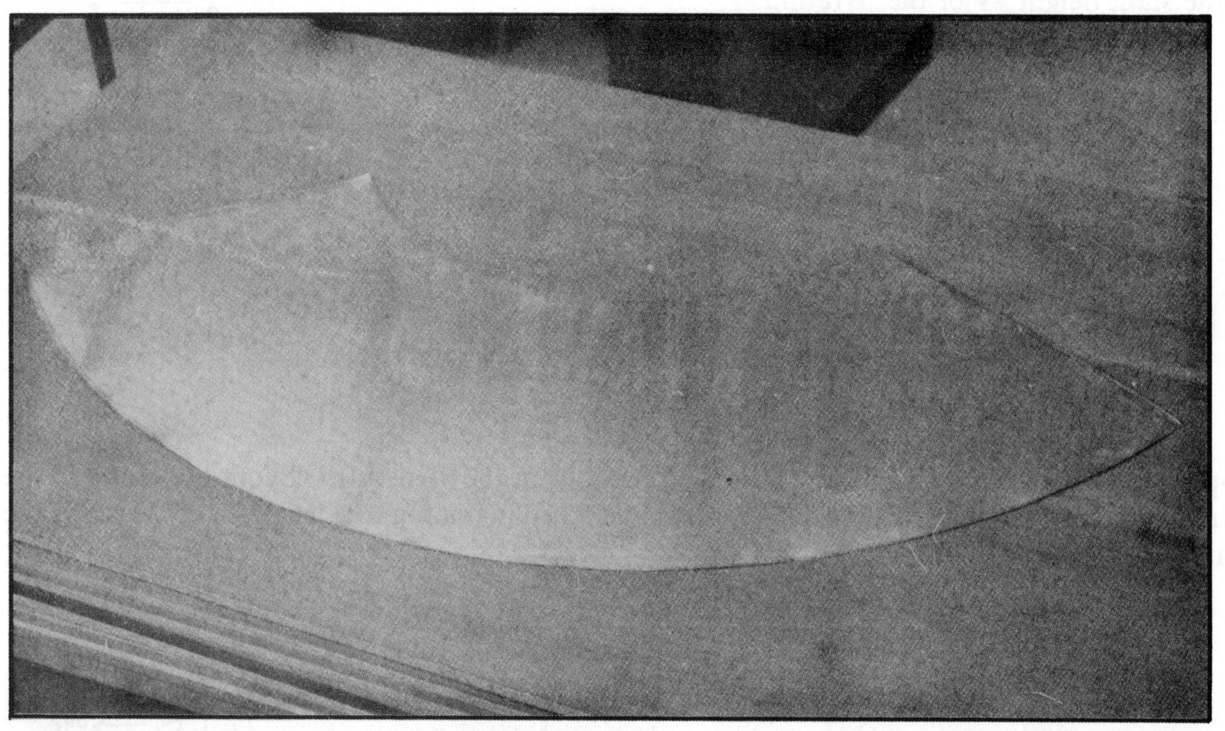

Unit 12
Enlargement and Reduction of Sections and Shapes

Certain figures must be changed in size by a proportionate method in order to maintain a similar pattern.

1. Construct a proportionate rectangle to ABCD with a side ratio of 4:5. Break AB into 5 equal parts and join diagonally AC. Draw a parallel bc to BC from 4, and cd parallel to CD.

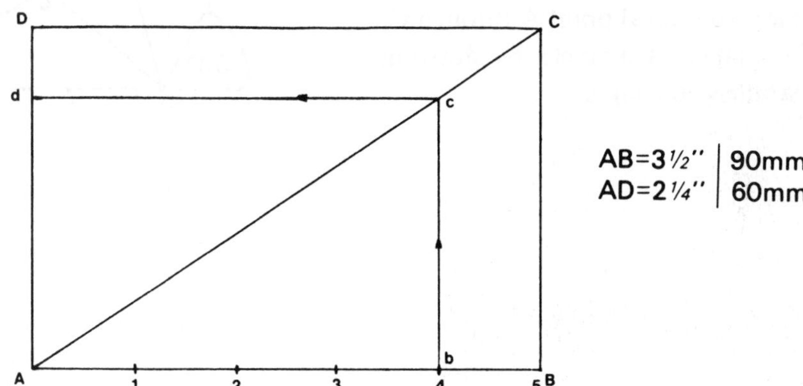

2. Construct a triangle with its sides increased to a ratio of 7:5. On a diagonal line from B, set out 7 equal divisions. Join 5 to c then lay off 7 parallel with 5c, This gives D when CB is extended. DE is now drawn parallel to Ac.

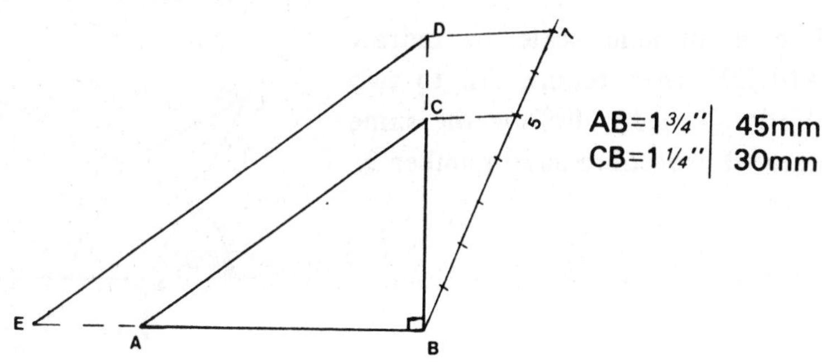

Polar Method

3. Using a suitable scale, draw figure ABCDE, then change proportionally the dimensions given to a ratio of 5:3. Divide line AB into 3 equal parts, extend AB and lay off 2 of these measured parts. Extend lines from focal point A through C, D, and E. Complete by drawing parallels starting at 5.

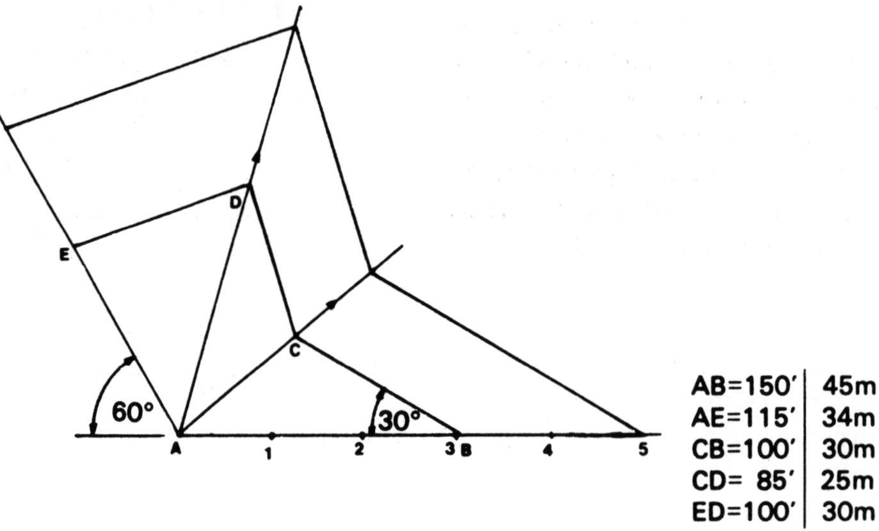

scale ⅛"=10'-0" | 1:1000

4. To decrease a figure in size proportionally.

Use a suitable scale to redraw ABCDE, then reduce AB to two thirds (20 feet). Follow the same parallel procedure as in number 3.

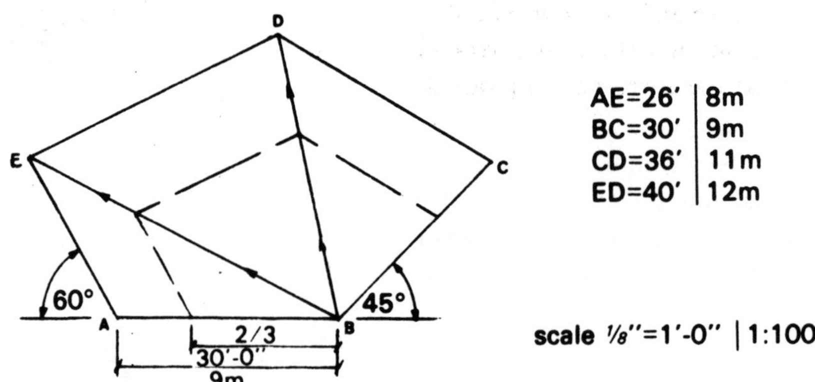

5. Reduce this stock moulding proportionally to half size. Given a new dimension Aa, follow the method in numbers 3 and 4. All parts of the moulding will then be reduced proportionally, thus keeping a perspective balance.

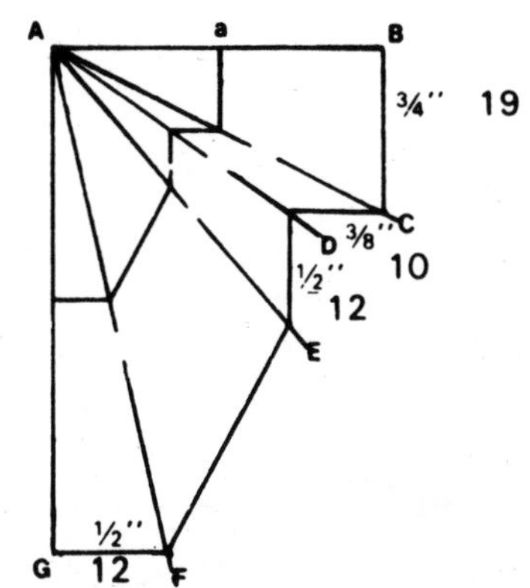

AB=1 ½"	40m
AG=2 ¼"	60mm
Aa=¾"	20mm

6. An optional method when working with sections is to select a focal point outside the figure, as at O. Radiate lines through all points where a change in direction occurs on the original. Make AG (the new dimension) parallel to ag. Draw AB parallel to ab and continue with BC, DC, and so on, letting each parallel cut off at the corresponding radial.

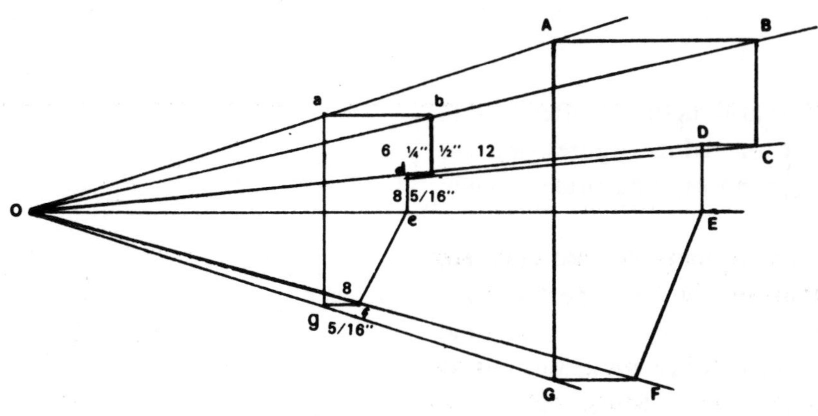

ab=¾"	20mm
ag=1 ¼"	30mm
AG=2 ¼"	60mm

7. This exercise is a reversal of number 6:

A typical crown moulding. When a curve has a radius, its transition is straight forward, but if the curves are not regular, ordinates would have to be plotted as in numbers 10 and 11.

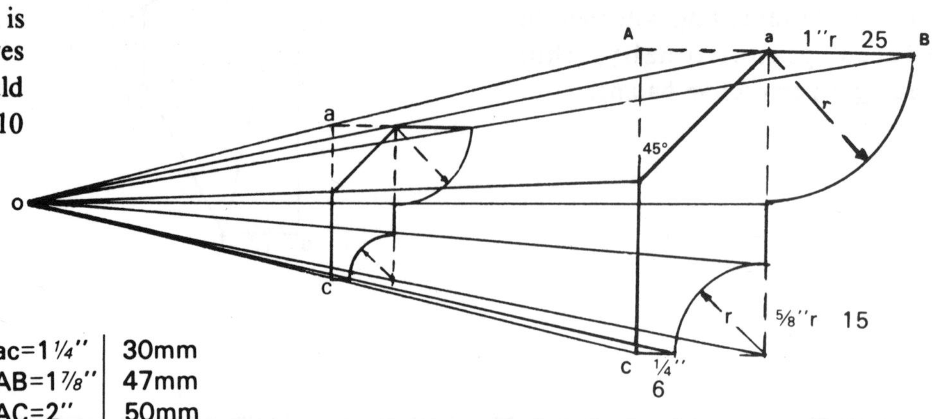

8. The grid method is often used to enlarge or reduce figures and for many purposes, is accurate enough:

Here a plot plan has to be enlarged, perhaps for a survey report. Draw a series of squares over the original (left), then redraw the same number of squares to the desired increase in size, in this case double. The copy can now be readily drawn by following positions on the squares. It is a good idea to number the squares.

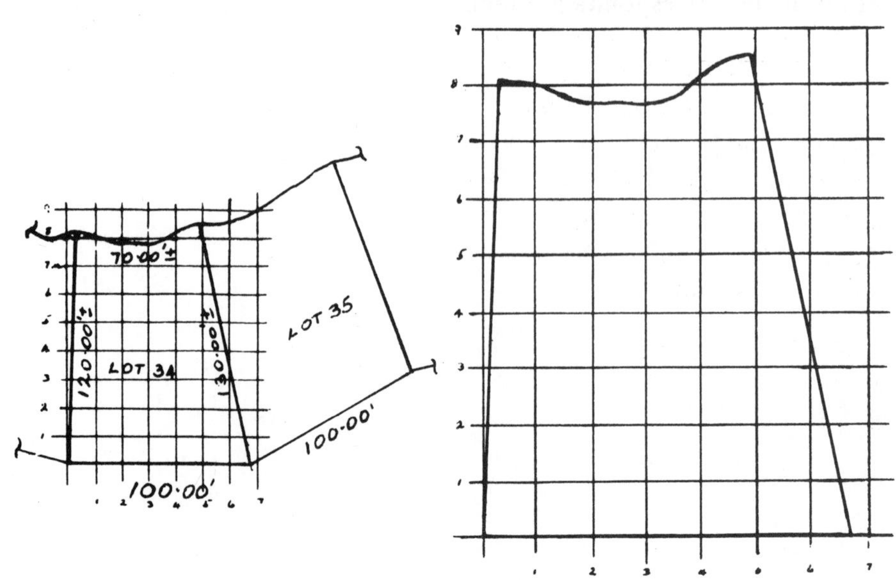

9. As you can see, this idea is useful to many artisans. The bird was reduced to 2/3.

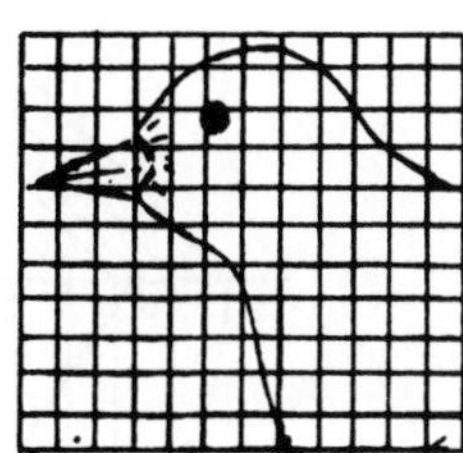

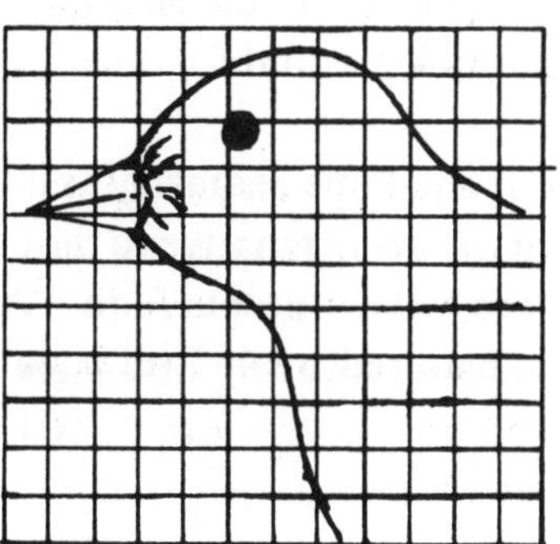

To increase or reduce a section in one direction only

10. Maintain the pattern, but reduce the width from 1¾" to 1¼". Layout the section ABCD. Select ordinates on the curve and extend all points a, b, c, d, e to BC. Produce AB so that BO is equal to the new dimension 1¼". Extend BC to C^1. Connect C to O and extend all points on BC to BO via parallels to CO. Using B as centre, swing all points on BO to BC^1 then extend them as perpendiculars to BC^1. The perpendiculars are then cut by vertical lines from abcde to make $a^1b^1c^1d^1e^1$. These points are now connected to show the reduced moulding section.

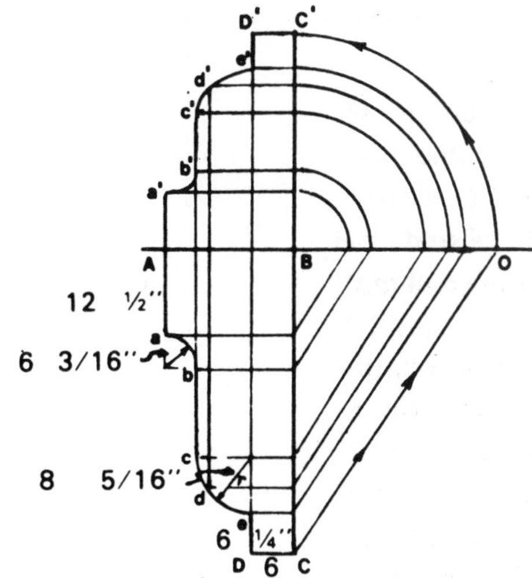

AB=¾"	19mm
BC=1¾"	45mm
BC¹=1¼"	30mm

11. Here the moulding is reduced in thickness only:

Use the same geometry as in number 10. As an alternative to the compass, here the points on AX are transferred by 45° set square.

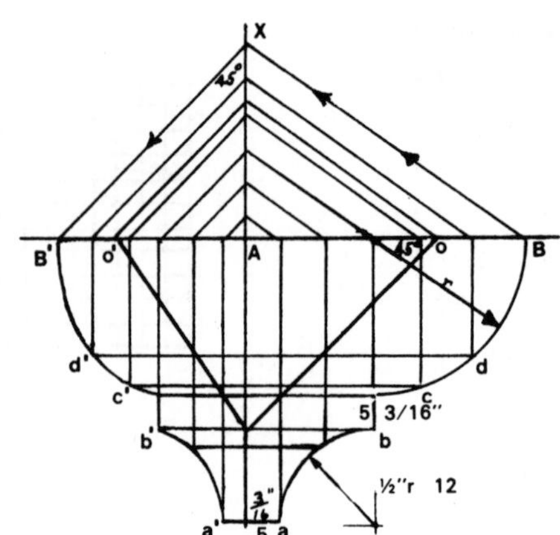

AB=1 ½"	40mm
AC=1 ½"	40mm
OB=½"	12mm
AB¹=1"	25mm

12. To increase or reduce to set dimensions of width and thickness:

Draw the original ABC. Extend AB to C¹ and AC to B¹. Connect B to B¹ and C to C¹, then a series of parallels and perpendiculars arrives at points of the new shape.

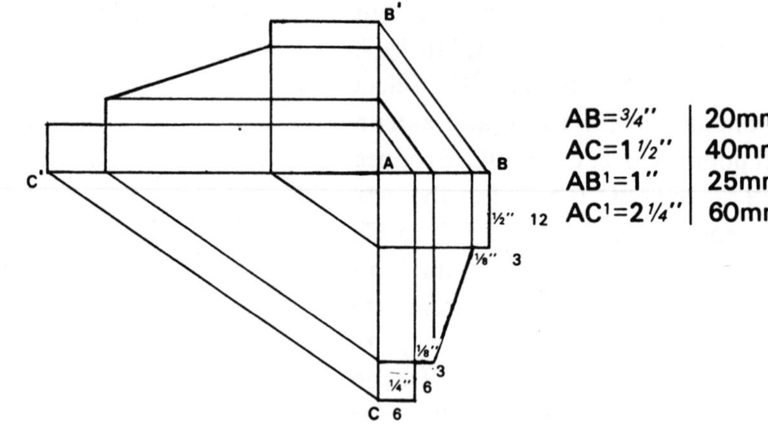

13. **Sometimes trim moulding has to be reduced in width. For example, the upper member of window casing is tight against a drop ceiling. The mitre WY of the two different widths is the diagonal of the rectangle WXYZ. The new section may be found by transfer of points via the mitre, or by the method shown in number 10.**

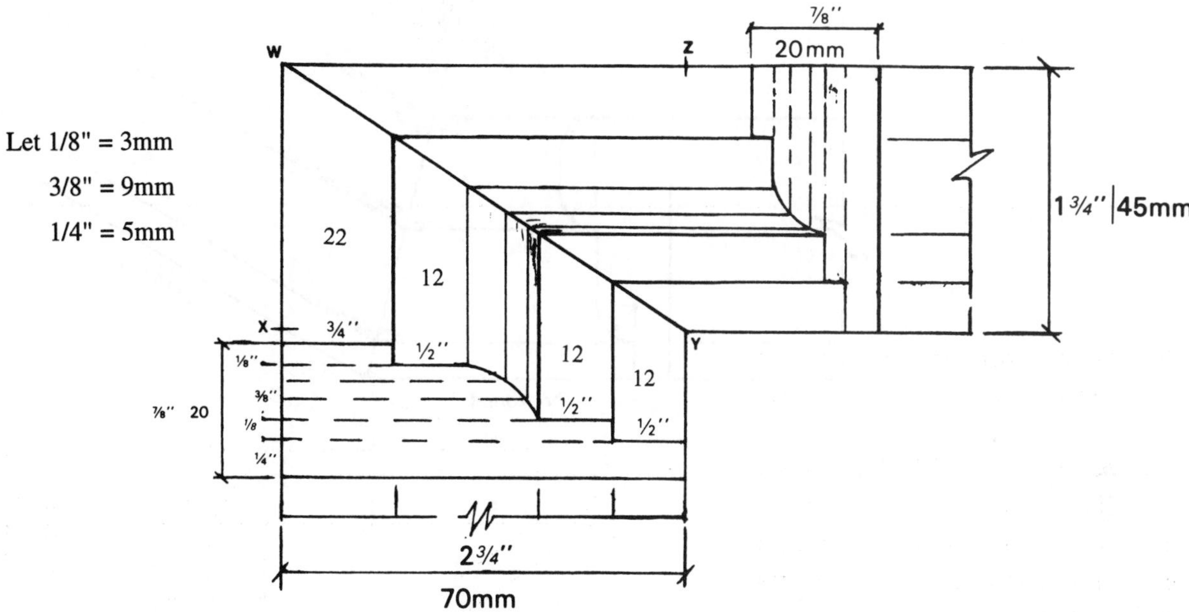

14. If trim starts to slope such as following a ramp or stairway, it is called a raked moulding. The section does not alter if the moulding stays on the same wall surface. The mitre cut is found by bisecting the angle AB.

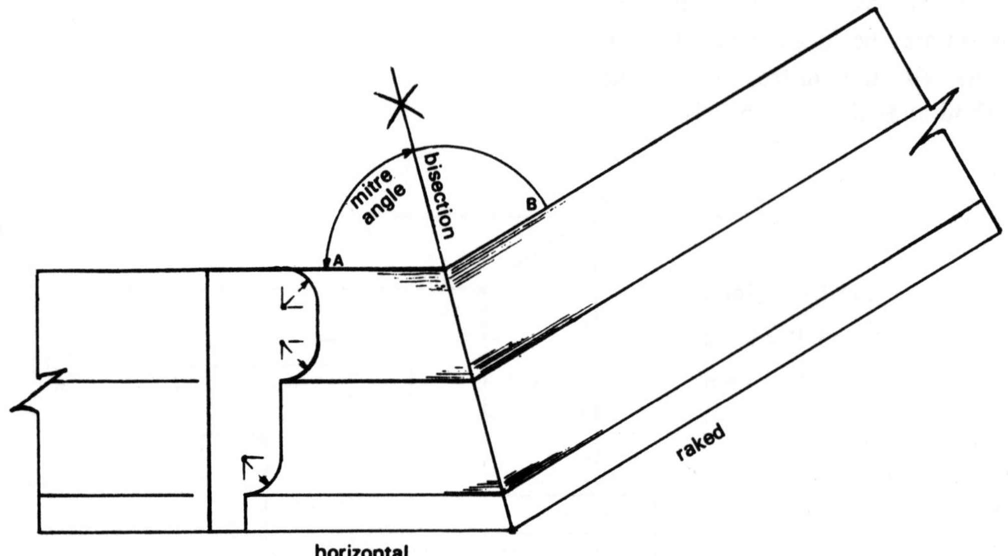

15. When a rake is applied on a different surface, the section changes:

Use the original in the most appropriate place, here it is on the incline. A new shape must now be discovered for the horizontal application. The plan and elevation are first drawn, then the original section is layed out as at XY. Select ordinates on the moulding and project to XY. Relocate these points on X¹Y¹. The points are further projected to meet at abcd and when properly connected, illustrate the desired shape. The mitre is a compound bevel much like a jack rafter. It is found by laying out the thickness of material around the plan angle, and the incline of rake. Apply these angles to a mitre box as shown. A is the plumb cut, B the side cut (top cut).

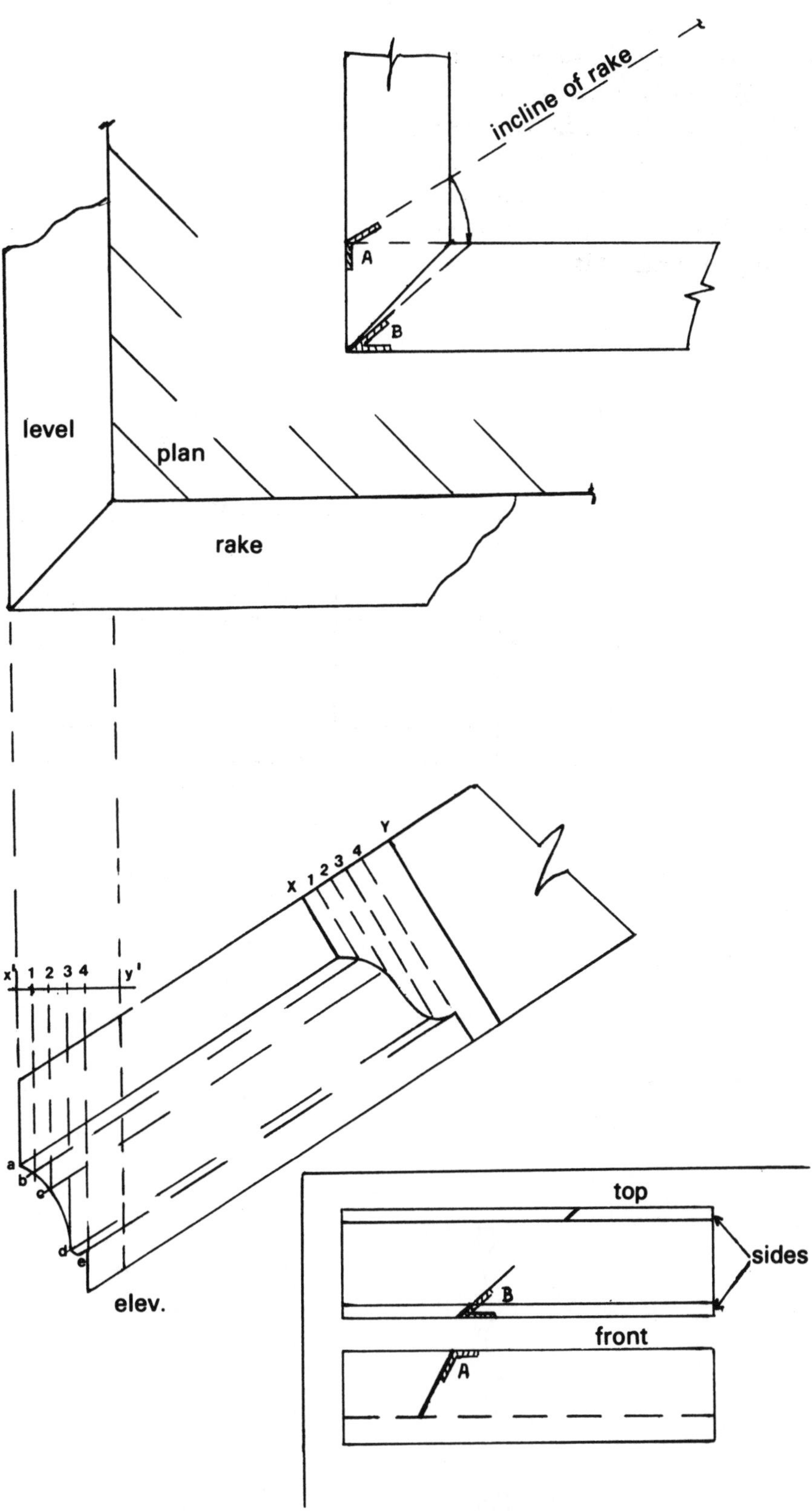

Most mouldings are based on the classic Greek and Roman designs. The following are samples of layout for some of them.

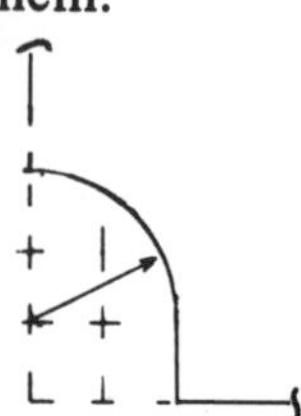

Cavetto

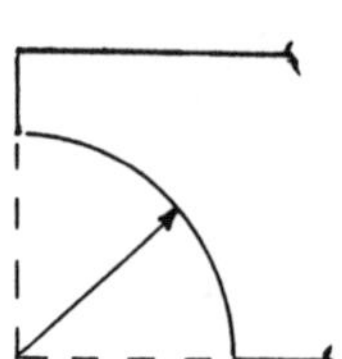

Cavetto

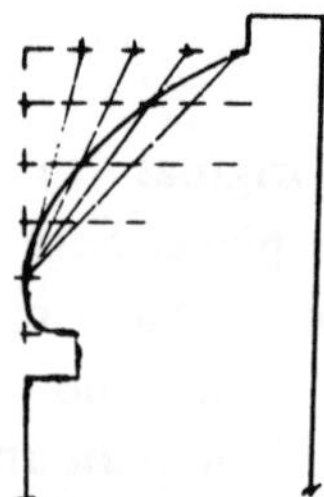

Parabolic Torus

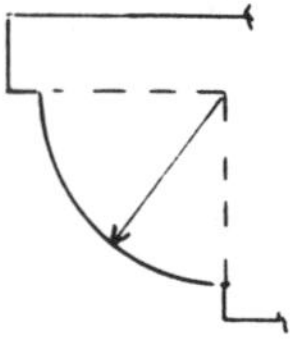

Ovolo

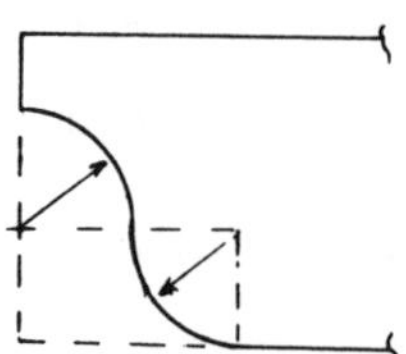

Cyma Recta

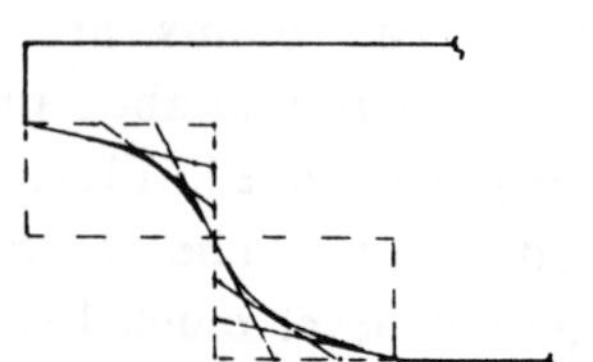

Elliptical

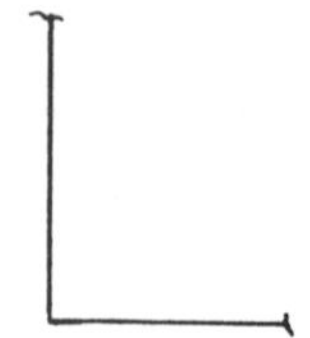

Square Arris

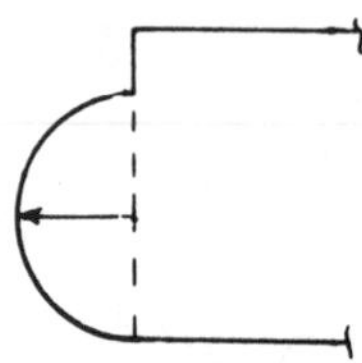

Torus

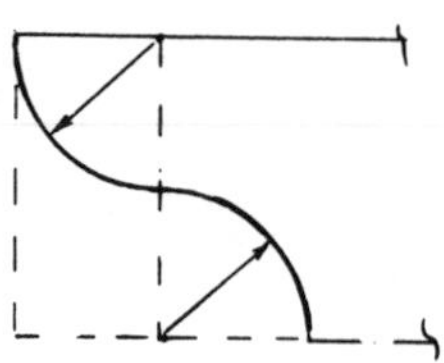

Cyma Reversa

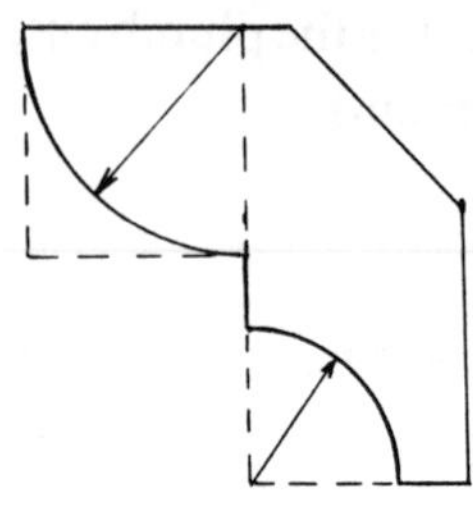

Ovolo and Cavetto

Reeding

Fluting

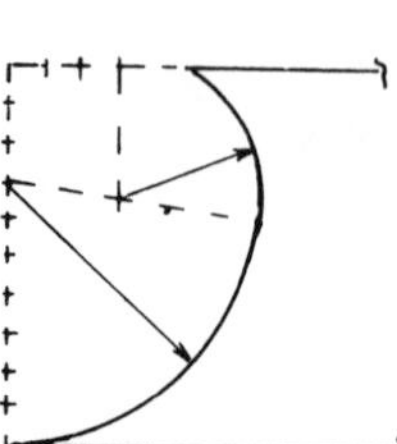

Scotia

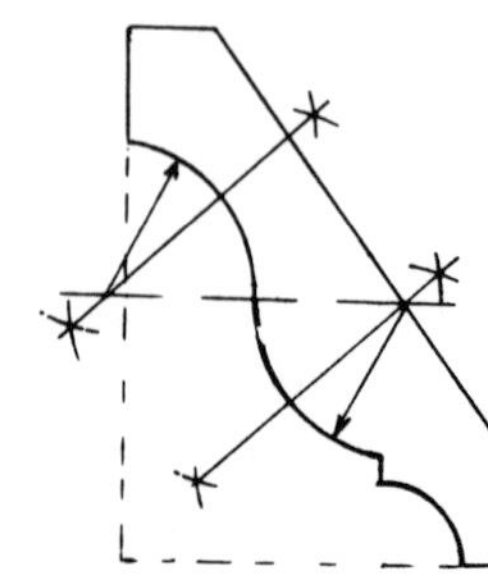

Cyma Recta or Cyma Reversa and Cavetto.

Discovery Exercises

1. Layout the six point star shown. Then show how it may be reduced by one third of the given dimension.

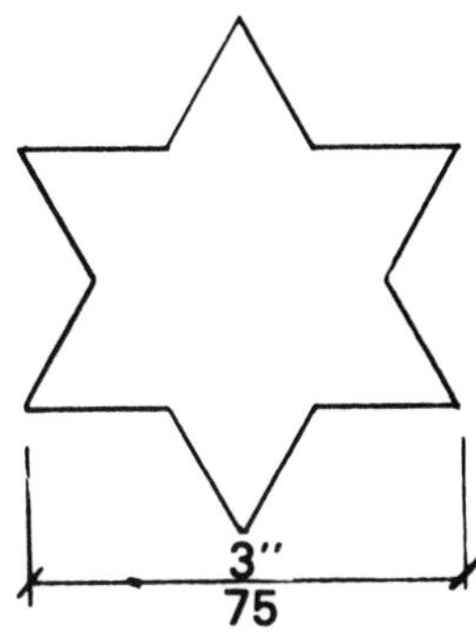

2. Enlarge proportionately the moulded baseboard section to 5" (130) wide. The thickness must remain the same.

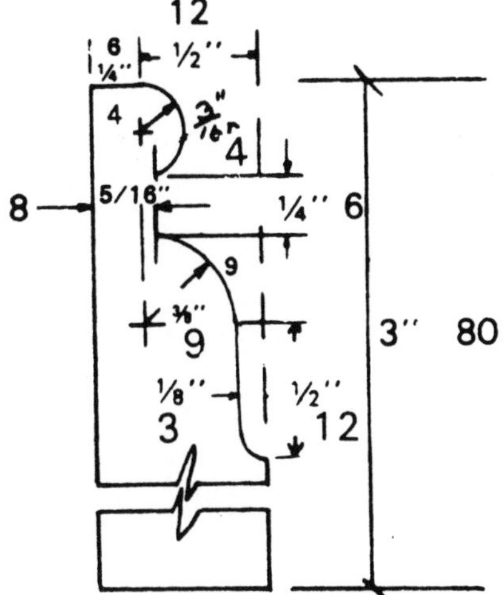

3. Reduce in proportion the illustrated bracket to 2" (50) wide and 4" (100) long.

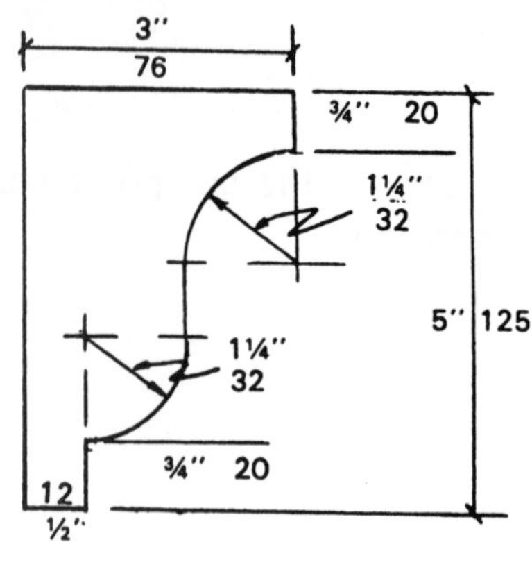

4. The given crown moulding section must be changed in proportion several times to suit job locations.

(a) reduce the section to 50% of the original
(b) reduce the thickness only, to 2½" (65) uv
(c) enlarge proportionately to 5" (130) x 5" (130)

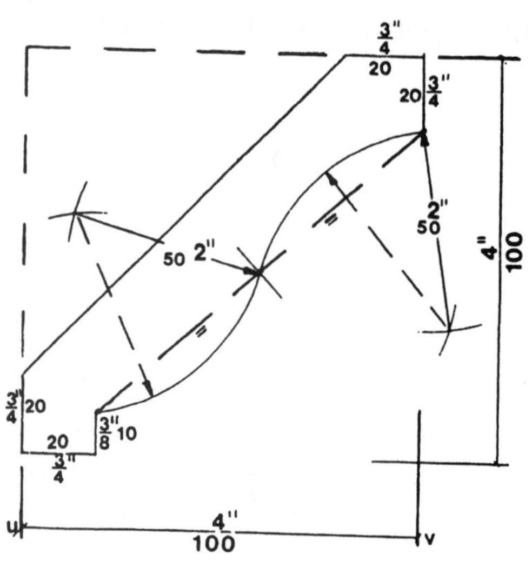

Useful Data

CONVERSION FACTORS FOR UNITS USED IN THE CONSTRUCTION INDUSTRY

(Conversion factors are taken to six significant figures where appropriate.)

	Metric to Imperial			Imperial to Metric		
Length	1 km	= 0.621 371	mile	1 mile	= 1.609 344	km
		= 49.7097	chain	1 chain	= 20.1168	m
	1 m	= 1.093 61	yd	1 yd	= 0.9144	m
		= 3.280 84	ft	1 ft	= 0.3048	m
	1 mm	= 0.039 370 1	in.		= 304.8	mm
				1 in.	= 25.4	mm
Area	1 km^2	= 0.386 102	$mile^2$	1 $mile^2$	= 2.589 99	km^2
	1 ha	= 2.471 05	acre	1 acre	= 0.404 686	ha
	1 m^2	= 1.195 99	yd^2		= 4046.86	m^2
		= 10.7639	ft^2	1 yd^2	= 0.836 127	m^2
	1 mm^2	= 0.001 550	$in.^2$	1 ft^2	= 0.092 903	m^2
				1 $in.^2$	= 645.16	mm^2
Volume, Capacity, Modulus of Section	1 m^3	= 0.810 713 x 10^{-3}	acre ft	1 acre ft	= 1233.48	m^3
		= 1.307 95	yd^3	1 yd^3	= 0.764 555	m^3
		= 35.3147	ft^3	1 ft^3	= 0.028 316 8	m^3
	1 mm^3	= 61.0237 x 10^{-6}	$in.^3$		= 28.3168	ℓ
	1 ℓ	= 0.035 314 7	ft^3	1 $in.^3$	= 16 387.1	mm^3
		= 0.219 969	gal		= 16.3871	ml
		= 1.759 76	pt	1 gal	= 4.546 09	ℓ
	1 ml	= 0.061 023 7	$in.^3$	1 pt	= 568.261	ml
		= 0.035 195 1	fl oz	1 fl oz	= 28.413 0	ml
Mass	1 tonne (t)	= 0.984 207	long ton	1 long ton	= 1.016 05	t
		= 19.684 1	cwt	1 cwt	= 50.8023	kg
	1 kg	= 2.204 62	lb	1 lb	= 0.453 592	kg
	1 g	= 0.035 274	oz	1 oz	= 28.3495	g
Mass/Unit Length	1 kg/m	= 0.671 969	lb/ft	1 lb/ft	= 1.488 16	kg/m
	1 g/m	= 3.547 99	lb/mile	1 lb/mile	= 0.281 849	g/m
Mass/Unit Area	1 kg/m^2	= 0.204 816	lb/ft^2	1 lb/ft^2	= 4.882 43	kg/m^2
	1 g/m^2	= 0.029 494	oz/yd^2	1 oz/yd^2	= 33.9057	g/m^2
		= 0.003 277 06	oz/ft^2	1 oz/ft^2	= 305.152	g/m^2
Density (Mass/Unit Volume)	1 kg/m^3	= 0.062 428	lb/ft^3	1 lb/ft^3	= 16.0185	kg/m^3
		= 1.685 56	lb/yd^3	1 lb/yd^3	= 0.593 278	kg/m^3
	1 t/m^3	= 0.752 48	long ton/yd^3	1 long ton/yd^3	= 1.328 94	t/m^3
Equivalent Temperature Value (°C = K-273.15)	°C = 5/9 °F-32			°F = 9/5 °C+32		

Useful Data

LENGTH

Inches and fractions of inches (up to 12 in.) to millimetres
Basis: 1 in. = 25.4 mm exactly

inches	0	1	2	3	4	5	6	7	8	9	10	11
fractions of inch	millimetres (mm)											
-	-	25.40	50.80	76.20	101.60	127.00	152.40	177.80	203.20	228.60	254.00	279.40
1/32	0.79	26.19	51.59	76.99	102.39	127.79	153.19	178.59	203.99	229.39	254.79	280.19
1/16	1.59	26.99	52.39	77.79	103.19	128.59	153.99	179.39	204.79	230.19	255.59	280.99
3/32	2.38	27.78	53.18	78.58	103.98	129.38	154.78	180.18	205.58	230.98	256.38	281.78
1/8	3.18	28.58	53.98	79.38	104.78	130.18	155.58	180.98	206.38	231.78	257.18	282.58
5/32	3.97	29.37	54.77	80.17	105.57	130.97	156.37	181.77	207.17	232.57	257.97	283.37
3/16	4.76	30.16	55.56	80.96	106.36	131.76	157.16	182.56	207.96	233.36	258.76	284.16
7/32	5.56	30.96	56.36	81.76	107.16	132.56	157.96	183.36	208.76	234.16	259.56	284.96
1/4	6.35	31.75	57.15	82.55	107.95	133.35	158.75	184.15	209.55	234.95	260.35	285.75
9/32	7.14	32.54	57.94	83.34	108.74	134.14	159.54	184.94	210.34	235.74	261.14	286.54
5/16	7.94	33.34	58.74	84.14	109.54	134.94	160.34	185.74	211.14	236.54	261.94	287.34
11/32	8.73	34.13	59.53	84.93	110.33	135.73	161.13	186.53	211.93	237.33	262.73	288.13
3/8	9.53	34.93	60.33	85.73	111.13	136.53	161.93	187.33	212.73	238.13	263.53	288.93
13/32	10.32	35.72	61.12	86.52	111.92	137.32	162.72	188.12	213.52	238.92	264.32	289.72
7/16	11.11	36.51	61.91	87.31	112.71	138.11	163.51	188.91	214.31	239.71	265.11	290.51
15/32	11.91	37.31	62.71	88.11	113.51	138.91	164.31	189.71	215.11	240.51	265.91	291.31
1/2	12.70	38.10	63.50	88.90	114.30	139.70	165.10	190.50	215.90	241.30	266.70	292.10
17/32	13.49	38.89	64.29	89.69	115.09	140.49	165.89	191.29	216.69	242.09	267.49	292.89
9/16	14.29	39.69	65.09	90.49	115.89	141.29	166.69	192.09	217.49	242.89	268.29	293.69
19/32	15.08	40.48	65.88	91.28	116.68	142.08	167.48	192.88	218.28	243.68	269.08	294.48
5/8	15.88	41.28	66.68	92.08	117.48	142.88	168.28	193.68	219.08	244.48	269.88	295.28
21/32	16.67	42.07	67.47	92.87	118.27	143.67	169.07	194.47	219.87	245.27	270.67	296.07
11/16	17.46	42.86	68.26	93.66	119.06	144.46	169.86	195.26	220.66	246.06	271.46	296.86
23/32	18.26	43.66	69.06	94.46	119.86	145.26	170.66	196.06	221.46	246.86	272.26	297.66
3/4	19.05	44.45	69.85	95.25	120.65	146.05	171.45	196.85	222.25	247.65	273.05	298.45
25/32	19.84	45.24	70.64	96.04	121.44	146.84	172.24	197.64	223.04	248.44	273.34	299.24
13/16	20.64	46.04	71.44	96.84	122.24	147.64	173.04	198.44	223.84	249.24	274.64	300.04
27/32	21.43	46.83	72.23	97.63	123.03	148.43	173.83	199.23	224.63	250.03	275.43	300.83
7/8	22.23	47.63	73.03	98.43	123.83	149.23	174.63	200.03	225.43	250.83	276.23	301.63
29/32	23.02	48.42	73.82	99.22	124.62	150.02	175.42	200.82	226.22	251.62	277.02	302.42
15/16	23.81	49.21	74.61	100.01	125.41	150.81	176.21	201.61	227.01	252.41	277.81	303.21
31/32	24.61	50.01	75.41	100.81	126.21	151.61	177.01	202.41	227.81	253.21	278.61	304.01

LENGTH
Feet and inches (up to 65 ft 8 in.) to metres - to three places of decimals
Basis: 1 ft = 0.3048 m exactly

feet	inches											
	0	1	2	3	4	5	6	7	8	9	10	11
	metres (m) - to three decimal places											
0	-	0.025	0.051	0.076	0.102	0.127	0.152	0.178	0.203	0.229	0.254	0.279
1	0.305	0.330	0.356	0.381	0.406	0.432	0.457	0.483	0.508	0.533	0.559	0.584
2	0.610	0.635	0.660	0.686	0.711	0.737	0.762	0.787	0.813	0.838	0.864	0.889
3	0.914	0.940	0.965	0.991	1.016	1.041	1.067	1.092	1.118	1.143	1.168	1.194
4	1.219	1.245	1.270	1.295	1.321	1.346	1.372	1.397	1.422	1.448	1.473	1.499
5	1.524	1.549	1.575	1.600	1.626	1.651	1.676	1.702	1.727	1.753	1.778	1.803
6	1.829	1.854	1.880	1.905	1.930	1.956	1.981	2.007	2.032	2.057	2.083	2.108
7	2.134	2.159	2.184	2.210	2.235	2.261	2.286	2.311	2.337	2.362	2.388	2.413
8	2.438	2.464	2.489	2.515	2.540	2.565	2.591	2.616	2.642	2.667	2.692	2.718
9	2.743	2.769	2.794	2.819	2.845	2.870	2.896	2.921	2.946	2.972	2.997	3.023
10	3.048	3.073	3.099	3.124	3.150	3.175	3.200	3.226	3.251	3.277	3.302	3.327
11	3.353	3.378	3.404	3.429	3.454	3.480	3.505	3.531	3.556	3.581	3.607	3.632
12	3.658	3.683	3.708	3.734	3.759	3.785	3.810	3.835	3.861	3.886	3.912	3.937
13	3.962	3.988	4.013	4.039	4.064	4.089	4.115	4.140	4.166	4.191	4.216	4.242
14	4.267	4.293	4.318	4.343	4.369	4.394	4.420	4.445	4.470	4.496	4.521	4.547
15	4.572	4.597	4.623	4.648	4.674	4.699	4.724	4.750	4.775	4.801	4.826	4.851
16	4.877	4.902	4.928	4.953	4.978	5.004	5.029	5.055	5.080	5.105	5.131	5.156
17	5.182	5.207	5.232	5.258	5.283	5.309	5.334	5.359	5.385	5.410	5.436	5.461
18	5.486	5.512	5.537	5.563	5.588	5.613	5.639	5.664	5.690	5.715	5.740	5.766
19	5.791	5.817	5.842	5.867	5.893	5.918	5.944	5.969	5.994	6.020	6.045	6.071
20	6.096	6.121	6.147	6.172	6.198	6.223	6.248	6.274	6.299	6.325	6.350	6.375
21	6.401	6.426	6.452	6.477	6.502	6.528	6.553	6.579	6.604	6.629	6.655	6.680
22	6.706	6.731	6.756	6.782	6.807	6.833	6.858	6.883	6.909	6.934	6.960	6.985
23	7.010	7.036	7.061	7.087	7.112	7.137	7.163	7.188	7.214	7.239	7.264	7.290
24	7.315	7.341	7.366	7.391	7.417	7.442	7.468	7.493	7.518	7.544	7.569	7.595
25	7.620	7.645	7.671	7.696	7.722	7.747	7.772	7.798	7.823	7.849	7.874	7.899
26	7.925	7.950	7.976	8.001	8.026	8.052	8.077	8.103	8.128	8.153	8.179	8.204
27	8.230	8.255	8.280	8.306	8.331	8.357	8.382	8.407	8.433	8.458	8.484	8.509
28	8.534	8.560	8.585	8.611	8.636	8.661	8.687	8.712	8.738	8.763	8.788	8.814
29	8.839	8.865	8.890	8.915	8.941	8.966	8.992	9.017	9.042	9.068	9.093	9.119
30	9.144	9.169	9.195	9.220	9.246	9.271	9.296	9.322	9.347	9.373	9.398	9.423
31	9.449	9.474	9.500	9.525	9.550	9.576	9.601	9.627	9.652	9.677	9.703	9.723
32	9.754	9.779	9.804	9.830	9.855	9.881	9.906	9.931	9.957	9.982	10.008	10.033
33	10.058	10.084	10.109	10.135	10.160	10.185	10.211	10.236	10.262	10.287	10.312	10.338
34	10.363	10.389	10.414	10.439	10.465	10.490	10.516	10.541	10.566	10.592	10.617	10.643
35	10.668	10.693	10.719	10.744	10.770	10.795	10.820	10.846	10.871	10.897	10.922	10.947
36	10.973	10.998	11.024	11.049	11.074	11.100	11.125	11.151	11.176	11.201	11.227	11.252
37	11.278	11.303	11.328	11.354	11.379	11.405	11.430	11.455	11.481	11.506	11.532	11.557
38	11.582	11.608	11.633	11.659	11.684	11.709	11.735	11.760	11.786	11.811	11.836	11.862
39	11.887	11.913	11.938	11.963	11.989	12.014	12.040	12.065	12.090	12.116	12.141	12.167
40	12.192	12.217	12.243	12.268	12.294	12.319	12.344	12.370	12.395	12.421	12.446	12.471
41	12.497	12.522	12.548	12.573	12.598	12.624	12.649	12.675	12.700	12.725	12.751	12.776
42	12.802	12.827	12.852	12.878	12.903	12.929	12.954	12.979	13.005	13.030	13.056	13.081
43	13.106	13.132	13.157	13.183	13.208	13.233	13.259	13.284	13.310	13.335	13.360	13.386
44	13.411	13.437	13.462	13.487	13.513	13.538	13.564	13.589	13.614	13.640	13.665	13.691
45	13.716	13.741	13.767	13.792	13.818	13.843	13.868	13.894	13.919	13.945	13.970	13.995
46	14.021	14.046	14.072	14.097	14.122	14.148	14.173	14.199	14.224	14.249	14.275	14.300
47	14.326	14.351	14.376	14.402	14.427	14.453	14.478	14.503	14.529	14.554	14.580	14.605
48	14.630	14.656	14.681	14.707	14.732	14.757	14.783	14.808	14.834	14.859	14.884	14.910
49	14.935	14.961	14.986	15.011	15.037	15.062	15.088	15.113	15.138	15.164	15.189	15.215
50	15.240	15.265	15.291	15.316	15.342	15.367	15.392	15.418	15.443	15.469	15.494	15.519
51	15.545	15.570	15.596	15.621	15.646	15.672	15.697	15.723	15.748	15.773	15.799	15.824
52	15.850	15.875	15.900	15.926	15.951	15.977	16.002	16.027	16.053	16.078	16.104	16.129
53	16.154	16.180	16.205	16.231	16.256	16.281	16.307	16.332	16.358	16.383	16.408	16.434
54	15.459	16.485	16.510	16.535	16.561	16.586	16.612	16.637	16.662	16.688	16.713	16.739
55	16.764	16.789	16.815	16.840	16.866	16.891	16.916	16.942	16.967	16.993	17.018	17.043
56	17.069	17.094	17.120	17.145	17.170	17.196	17.221	17.247	17.272	17.297	17.323	17.348
57	17.374	17.399	17.424	17.450	17.475	17.501	17.526	17.551	17.577	17.602	17.628	17.653
58	17.678	17.704	17.729	17.755	17.780	17.805	17.830	17.856	17.882	17.907	17.932	17.958
59	17.983	18.009	18.034	18.059	18.085	18.110	18.136	18.161	18.186	18.212	18.237	18.263
60	18.288	18.313	18.339	18.364	18.390	18.415	18.440	18.466	18.491	18.517	18.542	18.567
61	18.593	18.618	18.644	18.669	18.694	18.720	18.745	18.771	18.796	18.821	18.847	18.872
62	18.898	18.923	18.948	18.974	18.999	19.025	19.050	19.075	19.101	19.126	19.152	19.177
63	19.202	19.228	19.253	19.279	19.304	19.329	19.355	19.380	19.406	19.431	19.456	19.482
64	19.507	19.533	19.558	19.583	19.609	19.634	19.660	19.685	19.710	19.736	19.761	19.787
65	19.812	19.837	19.863	19.888	19.914	19.939	19.964	19.990	20.015			

AREA

Square inches (up to 100 in.^{2}) to square millimetres

Basis: 1 in.^{2} = 645.16 mm^2 exactly

square inches (in^2)	0	1	2	3	4	5	6	7	8	9
	square millimetres (mm^2)									
0	-	645.2	1 290.3	1 935.5	2 580.6	3 225.8	3 871.0	4 516.1	5 161.3	5 806.4
10	6 451.6	7 096.8	7 741.9	8 387.1	9 032.2	9 677.4	10 322.6	10 967.7	11 612.9	12 258.0
20	12 903.2	13 548.4	14 193.5	14 838.7	15 483.8	16 129.0	16 774.2	17 419.3	18 064.5	18 709.6
30	19 354.8	20 000.0	20 645.1	21 290.3	21 935.4	22 580.6	23 225.8	23 870.9	24 516.1	25 161.2
40	25 806.4	26 451.6	27 096.7	27 741.9	28 387:0	29 032.2	29 677.4	30 322.5	30 967.7	31 612.8
50	32 258.0	32 903.2	33 548.3	34 193.5	34 838.6	35 483.8	36 129.0	36 774.1	37 419.3	38 064.4
60	38 709.6	39 354.8	39 999.9	40 645.1	41 290.2	41 935.4	42 580.6	43 225.7	43 870.9	44 516.0
70	45 161.2	45 806.4	46 451.5	47 096.7	47 741.8	48 387.0	49 032.2	49 677.3	50 322.5	50 967.6
80	51 612.8	52 258.0	52 903.1	53 548.3	54 193.4	54 838.6	55 483.8	56 128.9	56 774.1	57 419.2
90	58 064.4	58 709.6	59 354.7	59 999.9	60 645.0	61 290.2	61 935.4	62 580.5	63 225.7	63 870.8
100	64 516.0									

Square feet (up to 400 ft^2) to square metres

Basis: 1 ft^2 = 0.092 903 m^2

square feet (ft^2)	0	1	2	3	4	5	6	7	8	9
	square metres (m^2)									
0	-	0.09	0.19	0.28	0.37	0.46	0.56	0.65	0.74	0.84
10	0.93	1.02	1.11	1.21	1.30	1.39	1.49	1.58	1.67	1.77
20	1.86	1.95	2.04	2.14	2.23	2.32	2.42	2.51	2.60	2.69
30	2.79	2.88	2.97	3.07	3.16	3.25	3.34	3.44	3.53	3.62
40	3.72	3.81	3.90	3.99	4.09	4.18	4.27	4.37	4.46	4.55
50	4.65	4.74	4.83	4.92	5.02	5.11	5.20	5.30	5.39	5.48
60	5.57	5.67	5.76	5.85	5.95	6.04	6.13	6.22	6.32	6.41
70	6.50	6.60	6.69	6.78	6.87	6.97	7.06	7.15	7.25	7.34
80	7.43	7.53	7.62	7.71	7.80	7.90	7.99	8.08	8.18	8.27
90	8.36	8.45	8.55	8.64	8.73	8.83	8.92	9.01	9.10	9.20
100	9.29	9.38	9.48	9.57	9.66	9.75	9.85	9.94	10.03	10.13
110	10.22	10.31	10.41	10.50	10.59	10.68	10.78	10.87	10.96	11.06
120	11.15	11.24	11.33	11.43	11.52	11.61	11.71	11.80	11.89	11.98
130	12.08	12.17	12.26	12.36	12.45	12.54	12.63	12.73	12.82	12.91
140	13.01	13.10	13.19	13.29	13.38	13.47	13.56	13.66	13.75	13.84
150	13.94	14.03	14.12	14.21	14.31	14.40	14.49	14.59	14.68	14.77
160	14.86	14.96	15.05	15.14	15.24	15.33	15.42	15.51	15.61	15.70
170	15.79	15.89	15.98	16.07	16.17	16.26	16.35	16.44	16.54	16.63
180	16.72	16.82	16.91	17.00	17.09	17.19	17.28	17.37	17.47	17.56
190	17.65	17.74	17.84	17.93	18.02	18.12	18.21	18.30	18.39	18.49
200	18.58	18.67	18.77	18.86	18.95	19.05	19.14	19.23	19.32	19.42
210	19.51	19.60	19.70	19.79	19.88	19.97	20.07	20.16	20.25	20.35
220	20.44	20.53	20.62	20.72	20.81	20.90	21.00	21.09	21.18	21.27
230	21.37	21.46	21.55	21.65	21.74	21.83	21.93	22.02	22.11	22.20
240	22.30	22.39	22.48	22.58	22.67	22.76	22.85	22.95	23.04	23.13
250	23.23	23.32	23.41	23.50	23.60	23.69	23.78	23.88	23.97	24.06
260	24.15	24.25	24.34	24.43	24.53	24.62	24.71	24.81	24.90	24.99
270	25.08	25.18	25.27	25.36	25.46	25.55	25.64	25.73	25.83	25.92
280	26.01	26.11	26.20	26.29	26.38	26.48	26.57	26.66	26.76	26.85
290	26.94	27.03	27.13	27.22	27.31	27.41	27.50	27.59	27.69	27.78
300	27.87	27.96	28.06	28.15	28.24	28.34	28.43	28.52	28.61	28.71
310	28.80	28.89	28.99	29.08	29.17	29.26	29.36	29.45	29.54	29.64
320	29.73	29.82	29.91	30.01	30.10	30.19	30.29	30.38	30.47	30.57
330	30.66	30.75	30.84	30.94	31.03	31.12	31.22	31.31	31.40	31.49
340	31.59	31.68	31.77	31.87	31.96	32.05	32.14	32.24	32.33	32.42
350	32.52	32.61	32.70	32.79	32.89	32.98	33.07	33.17	33.26	33.35
360	33.45	33.54	33.63	33.72	33.82	33.91	34.00	34.10	34.19	34.28
370	34.37	34.47	34.56	34.65	34.75	34.84	34.93	35.02	35.12	35.21
380	35.30	35.40	35.49	35.58	35.67	35.77	35.86	35.95	36.05	36.14
390	36.23	36.33	36.42	36.51	36.60	36.70	36.79	36.88	36.98	37.07
400	37.16									

VOLUME

Cubic feet (up to 106 ft^3) to cubic metres

Basis: 1 ft^3 = 0.028 317 m^3

cubic feet (ft^3)	0	1	2	3	4	5	6	7	8	9
	cubic metres (m^3)									
0	-	0.03	0.06	0.08	0.11	0.14	0.17	0.20	0.23	0.25
10	0.28	0.31	0.34	0.37	0.40	0.42	0.45	0.48	0.51	0.54
20	0.57	0.59	0.62	0.65	0.68	0.71	0.74	0.76	0.79	0.82
30	0.85	0.88	0.91	0.93	0.96	0.99	1.02	1.05	1.08	1.10
40	1.13	1.16	1.19	1.22	1.25	1.27	1.30	1.33	1.36	1.39
50	1.42	1.44	1.47	1.50	1.53	1.56	1.59	1.61	1.64	1.67
60	1.70	1.73	1.76	1.78	1.81	1.84	1.87	1.90	1.93	1.95
70	1.98	2.01	2.04	2.07	2.10	2.12	2.15	2.18	2.21	2.24
80	2.27	2.29	2.32	2.35	2.38	2.41	2.44	2.46	2.49	2.52
90	2.55	2.58	2.61	2.63	2.66	2.69	2.72	2.75	2.78	2.80
100	2.83	2.86	2.89	2.92	2.94	2.97	3.00			

Gallons (up to 110 gal) to litres

Basis: 1 gal = 4.546 09 ℓ

gallons	0	1	2	3	4	5	6	7	8	9
	litres (ℓ)									
0	-	4.55	9.09	13.64	18.18	22.73	27.28	31.82	36.37	40.91
10	45.46	50.01	54.55	59.10	63.65	68.19	72.74	77.28	81.83	86.38
20	90.92	95.47	100.01	104.56	109.11	113.65	118.20	122.74	127.29	131.84
30	136.38	140.93	145.48	150.02	154.57	159.11	163.66	168.21	172.75	177.30
40	181.84	186.39	190.94	195.48	200.03	204.57	209.12	213.67	218.21	222.76
50	227.30	231.85	236.40	240.94	245.49	250.03	254.58	259.13	263.67	268.22
60	272.77	277.31	281.86	286.40	290.95	295.50	300.04	304.59	309.13	313.68
70	318.23	322.77	327.32	331.87	336.41	340.96	345.50	350.05	354.60	359.14
80	363.69	368.23	372.78	377.33	381.87	386.42	390.96	395.51	400.06	404.60
90	409.15	413.69	418.24	422.79	427.33	431.88	436.42	440.97	445.52	450.06
100	454.61	459.16	463.70	468.25	472.79	477.34	481.89	486.43	490.98	495.52
110	500.07									

MASS

Pounds (up to 250 lb) to kg

Basis: 1 lb = 0.453 592 kg

pounds (lb)	0	1	2	3	4	5	6	7	8	9
	kilograms (kg)									
0	-	0.45	0.91	1.36	1.81	2.27	2.72	3.18	3.63	4.08
10	4.54	4.99	5.44	5.90	6.35	6.80	7.26	7.71	8.16	8.62
20	9.07	9.53	9.98	10.43	10.89	11.34	11.79	12.25	12.70	13.15
30	13.61	14.06	14.52	14.97	15.42	15.88	16.33	16.78	17.24	17.69
40	18.14	18.60	19.05	19.50	19.96	20.41	20.87	21.32	21.77	22.23
50	22.68	23.13	23.59	24.04	24.49	24.95	25.40	25.85	26.31	26.76
60	27.22	27.67	28.12	28.58	29.03	29.48	29.94	30.39	30.84	31.30
70	31.75	32.21	32.66	33.11	33.57	34.02	34.47	34.93	35.38	35.83
80	36.29	36.74	37.19	37.65	38.10	38.56	39.01	39.46	39.92	40.37
90	40.82	41.28	41.73	42.18	42.64	43.09	43.54	44.00	44.45	44.91
100	45.36	45.81	46.27	46.72	47.17	47.63	48.08	48.53	48.99	49.44
110	49.90	50.35	50.80	51.26	51.71	52.16	52.62	53.07	53.52	53.98
120	54.43	54.88	55.34	55.79	56.25	56.70	57.15	57.61	58.06	58.51
130	58.97	59.42	59.87	60.33	60.78	61.24	61.69	62.14	62.60	63.05
140	63.50	63.96	64.41	64.86	65.32	65.77	66.22	66.68	67.13	67.59
150	68.04	68.49	68.95	69.40	69.85	70.31	70.76	71.21	71.67	72.12
160	72.57	73.03	73.48	73.94	74.39	74.84	75.30	75.75	76.20	76.66
170	77.11	77.56	78.02	78.47	78.93	79.38	79.83	80.29	80.74	81.19
180	81.65	82.10	82.55	83.01	83.46	83.91	84.37	84.82	85.28	85.73
190	86.18	86.64	87.09	87.54	88.00	88.45	88.90	89.36	89.81	90.26
200	90.72	91.17	91.63	92.08	92.53	92.99	93.44	93.89	94.35	94.80
210	95.25	95.71	96.16	96.62	97.07	97.52	97.98	98.43	98.88	99.34
220	99.79	100.24	100.70	101.15	101.61	102.06	102.51	102.97	103.42	103.87
230	104.33	104.78	105.23	105.69	106.14	106.59	107.05	107.50	107.96	108.41
240	108.86	109.32	109.77	110.22	110.68	111.13	111.58	112.04	112.49	112.94
250	113.40									

SLOPE

Conventions

Previous practice used different ways to express slope in drawings: an angle was expressed in degrees and minutes; ratio of vertical to horizontal was expressed as percentages. The use of millimetres or metres in building drawings favours the adaptation of nondimensional ratio as the expression of slope. This has been recognized in the draft standard CSA B78.3 "Building Drawing Practics." For instance, a slope of 1:2 (ie. a vertical: horizontal ratio of 100:200 or of 50:100) is a clearer way of expressing slope than the angular equivalent of 26°34'. It also simplifies site layout. In special cases where a high degree of accuracy is required, however, the draft standard accepts angular expressions for slopes.

Expression of slope as a ratio

In expressing slope as a ratio, the vertical component is always shown first. Thus for slopes less than 45°, the first number should always be shown as unity, eg. a ratio of 1:5 indicates a rise of 1mm for every 5mm of horizontal dimension, or 1m for every 5m, etc. For slopes steeper that 45°, the second number, the horizontal component, should always be unity to facilitate easy verification. Ratio of 5:1 expresses a rise of 5mm for a horizontal dimension of 1mm or 5m for each 1m, etc. The use of mixed units, such as 1mm in 10m, 5m in 1km, should be avoided if possible.

The following table shows ratios, angular expressions, and percentages for easy calculation from one system to another.

From the *Manual on Metric Building Drawing Practice.*

Expression of Slope

Ratio ($\frac{y}{x}$)	Angle	Percentage (%)
Shallow Slopes		
1:100	0° 34'	1
1:67	0° 52'	1.5
1:57	1°	1.75
1:50	1° 09'	2
1:40	1° 26'	2.5
1:33	1° 43'	3
1:29	2°	3.5
1:25	2° 17'	4
1:20	2° 52'	5
1:19	3°	5.25
Slight Slopes		
1:17	3° 26'	6
1:15	3° 48'	6.7
1:14.3	4°	7
1:12	4° 46'	8.3
1:11.4	5°	8.75
1:10	5° 43'	10
1:9.5	6°	10.5
1:8	7° 07'	12.5
1:7.1	8°	14
1:6.7	8° 32'	15
1:6	9° 28'	16.7
1:5.7	10°	17.6
1:5	11° 19'	20
1:4.5	12° 30'	22.2
1:4	14° 02'	25
Medium Slopes		
1:3.7	15°	26.8
1:3.3	16° 42'	30
1:3	18° 26'	33.3
1:2.75	20°	36.4
1:2.5	21° 48'	40
1:2.4	22° 30'	41.4
1:2.15	25°	46.6
1:2	26° 34'	50
1:1.73	30°	57.5
1:1.67	30° 58'	60
1:1.5	33° 42'	67
1:1.33	36° 52'	75
1:1.2	40°	84
1:1	45°	100
Steep Slopes		
1.2:1	50°	119
1.43:1	55°	143
1.5:1	56° 19'	150
1.73:1	60°	173
2:1	63° 26'	200
2.15:1	65°	215
2.5:1	68° 12'	250
2.75:1	70°	275
3:1	71° 34'	300
3.73:1	75°	373
4:1	75° 58'	400
5:1	78° 42'	500
5.67:1	80°	567
6:1	80° 32'	600
11.43:1	85°	1143
∞	90°	∞

Appendix

Adapting the Ellipse to H Window Frame

An elliptical window, transom light or mirror frame is always aesthetically graceful. Many designers use the ellipse for decor in places such as a window over a stairwell, in a hallway or within a gable.

The frame may be cut from solid wood and rebated on a shaper or router. The joints may be open mortise and tenon or butted and secured with handrail bolts; however, an easier approach is to laminate.

First, lay out half the shape on cardboard or plywood for two templates. One template is required for the outer laminate, and one for the reduced (rebate side) inner laminate. Second, mark in the lines of axes for joints on the reduced template. Third, choose a location (according to width of available material for cutting the deeper curve) and lay out normals for the face joints. Normals are simply the bisection of lines from F_1 and F_2 to your selected points, as indicated in the basic layout at J. Finally, mark all pieces from the templates, cut and laminate together.

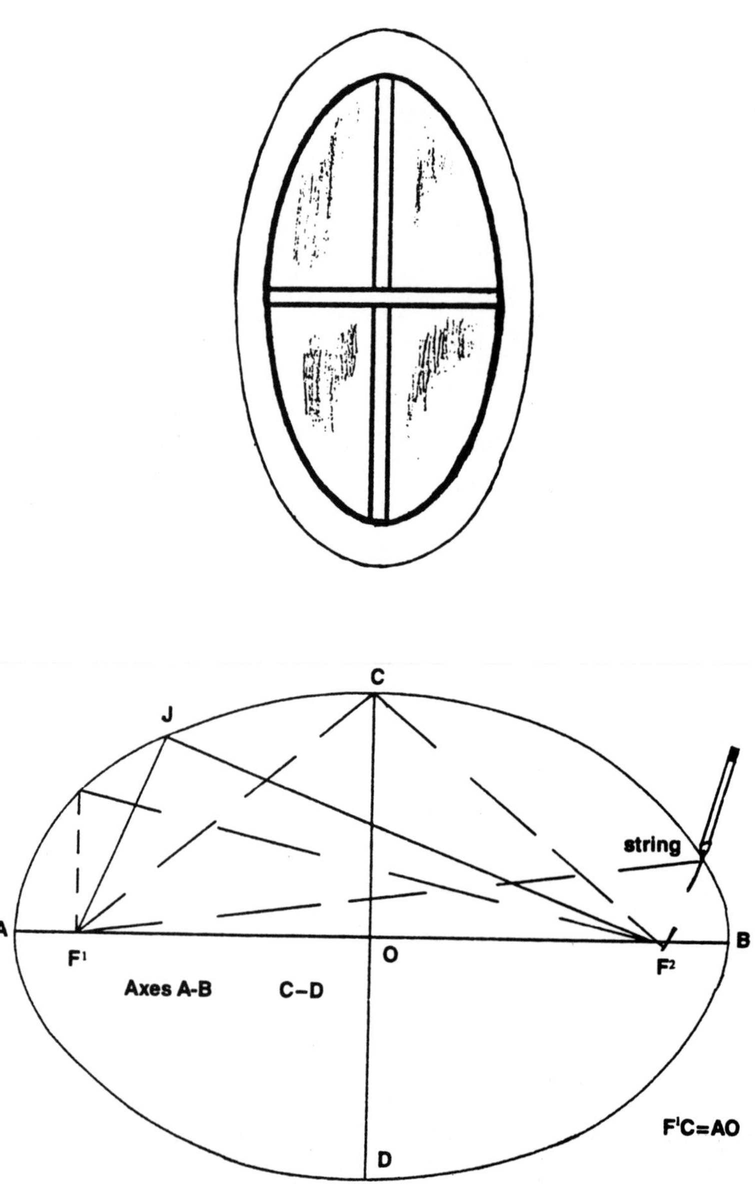

Run a router around the inner edge for a moulding, or apply a bead (as shown) with the sharpened head of a screw fixed into a narrow block and drawn around the frame.

The glass stop can be made of two strips of 1/8" plywood cut across the grain.

Method

First, lay out the axes. Then find the focus points by striking arcs at F1 and F2 using AO as a radius from C. A string is now tied at F1, stretched taut around C, and tied at F2. Remove pin C and insert a pencil. Keeping string taut, traverse to A and B.

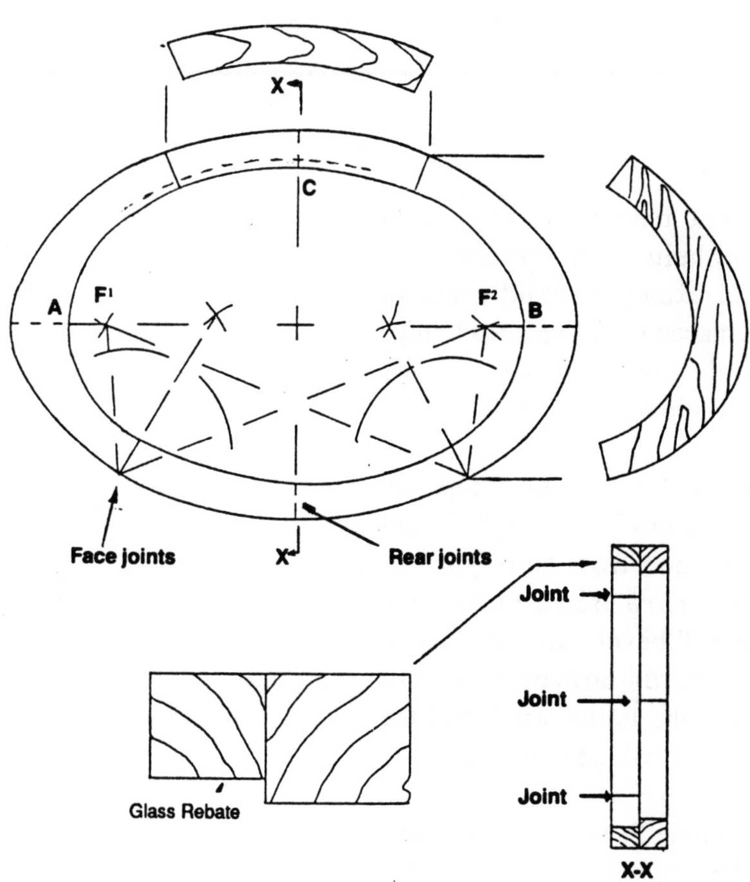

Renovation Geometry

Authentic renovations often need a little help from geometry. A typical example would occur in the replacement of splayed linings around the interior of the windows in the home.

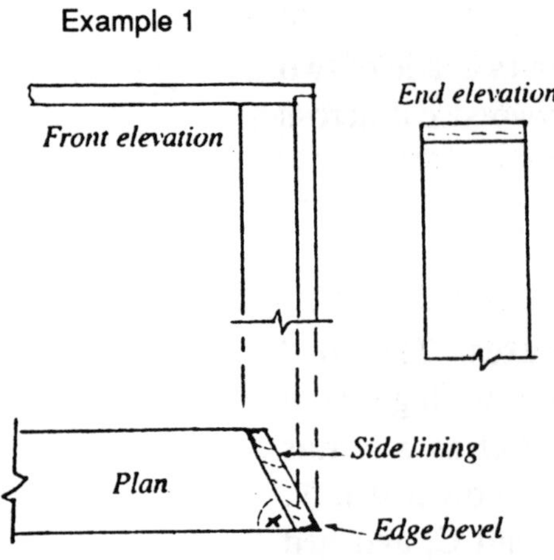

Example 1 has a level top and splayed sides. The angle X can be measured from an original or from window frame to trip line using a T bevel. Lay this angle across cardboard or plywood; measure the lining width and the edge bevel will be obvious.

In Example 2, both the sides and top slope require angle cuts at the joints. Lay out a plan and elevation, then using A as a centre, turn AB into the elevation. Where the projected top elevation line cuts a D, join D to C. This gives the face bevel for top and side linings provided their angle from the window is the same. An edge bevel can be seen in the first layout. A butt joint can also be used.

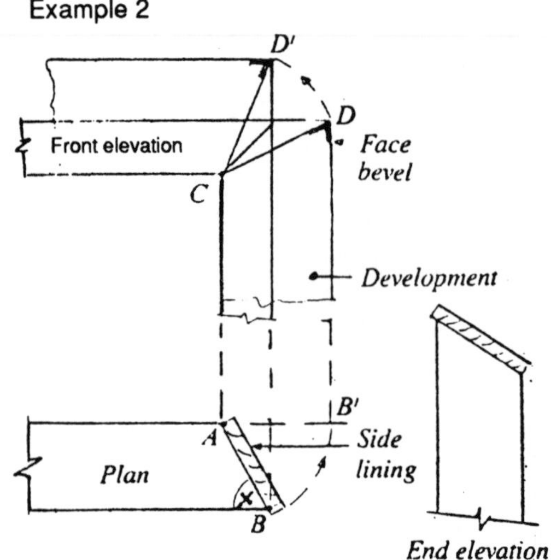

If a mitre is preferred, see Example 3. This work is similar to hopper layout.

1. The linings are splayed at 60°. Use 1 as centre a, to swing the lining face 1-2 into the elevation to meet a line produced from the head at 4.

2. Joining 4-5 give B the face bevel. The face edge bevel is seen at A, thus making 90° over that edge for a joint as seen in the elevation.

3. If a mitre is desired, strike a line anywhere across 5-6 at 90° to meet the edges at 0-0. Transfer 5-4 (joint length) to 6-7, cut by a 90° from 5. Now strike a tangent to make the dihedral — half of which is C the mitre.

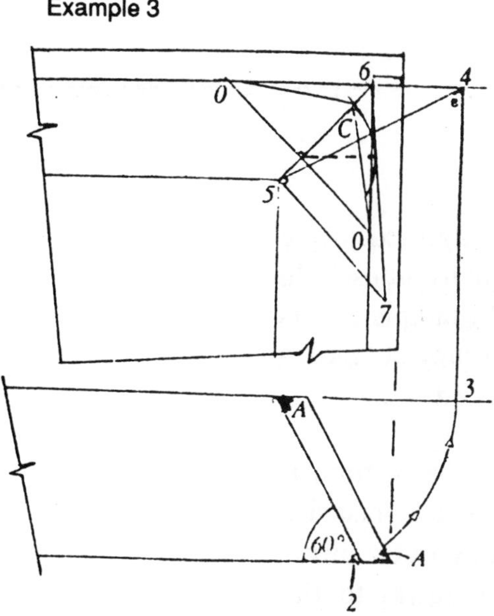

Planters with a Twist

Geometry can be applied to the construction of many household projects. A good example of this is a unique spring project: garden planters built from 2 x 4's.

Example 1

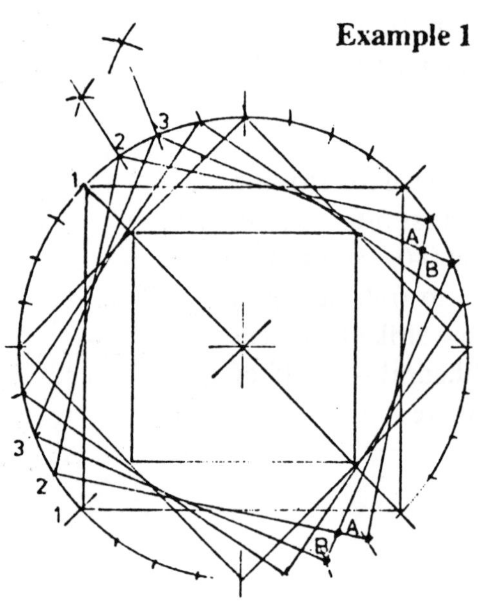

In Example 1, the basic fram is square; 2' seems to be a good size to work with. Lay out a circle with a diameter equal to the diagonal of the square, then divide the circle into 32 equal divisions. This may be done with a protractor or by bisection. These points represent the location of the corners of each layered square. The distances A and B should be remeasured on each alternate frame to position them in the build up of the box.

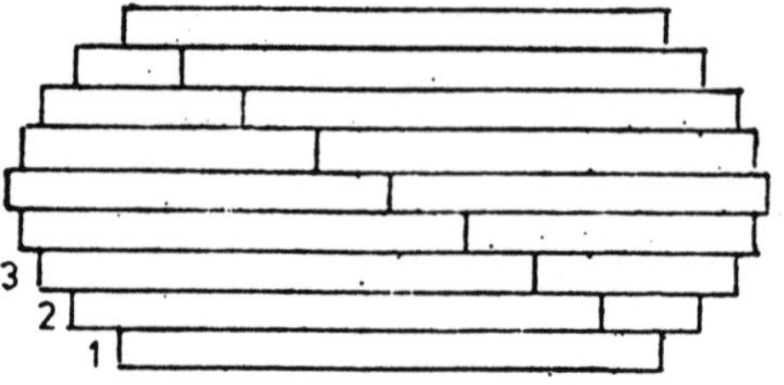

In Example 2, the equilateral triangle is used following the same procedure, but with 24 sector divisions. Again A and B equal the amount to offset each frame at the corners. With 9 frames layered, the planter would finish about 14" high with the top frame lining up with the base, thus demonstrating a good spiral. Should you wish to decrease the spiral, more points may be used on the layout circle.

For this project, I would suggest a mitre joint be used and cedar or treated spruce for the material.

Example 2

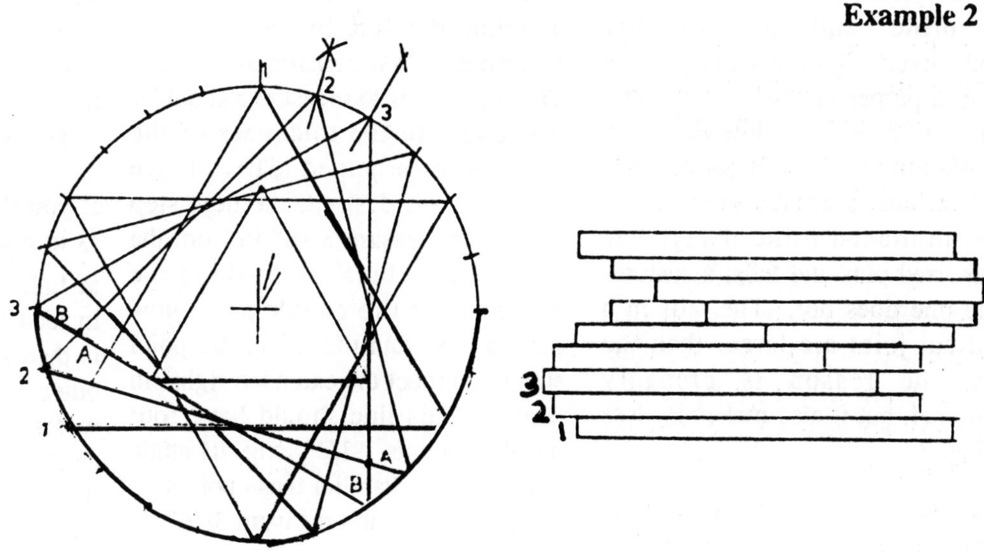

Dovetail Joints

Dovetail joints, when well made, are the pride and joy of any woodworker. Basic geometry is the key to a proper layout for this joint. The difference between a manufactured dovetail joint and a hand-cut one is usually obvious. A mass produced joint always has equal sized pins and tails; a custom-made one does not. The tails in a hand-cut joint are larger than the pins; the reason is probably traditional, but it also makes for less cutting.

The layout is based on the equal division of a line by parallels. In Example 1, pin centres are found. The first step is to measure a shoulder line equal to the thickness of the material to be joined (Th). If five 3/8" pins are desired, the next step will be to measure 3/16" in from the outer edges, thus establishing the centre for the two outside pins. Now draw a line from the left centre point sloping to meet the extended right pin centre. This line should have four equal segments 1-1/2" long; its angle is irrelevant. Parallel these points up to the top edge and measure 3/16" on each side.

To determine the slope on the pin sides, use a ratio of one in six for softwood or one in eight for hardwood. Lay out the appropriate slope and transfer with a T bevel.

If you desire equal size pins and tails as in Example 2, use a measuring line at half the thickness of stock TH/2 instead of the top edge. The division is then made on this line between the outer edges of the board to determine width of pins and tails rather than centres. Lay the appropriate slope through the points on this line.

Example 1

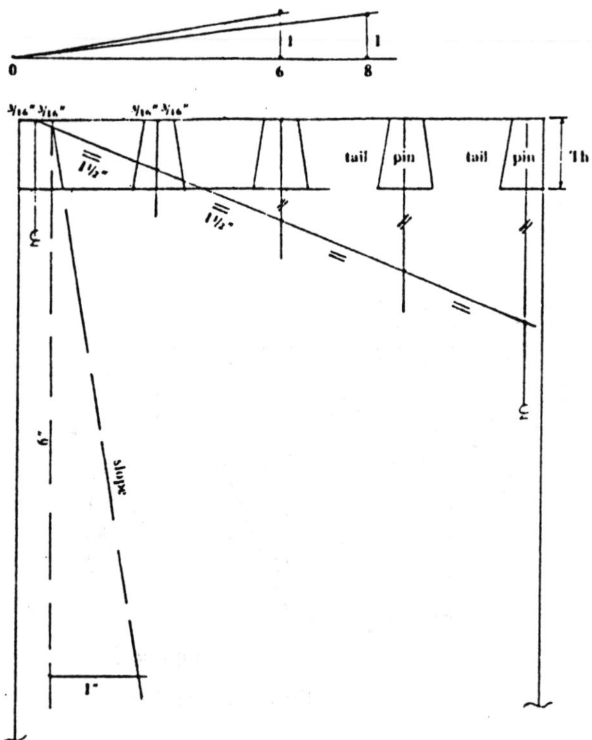

Example 2

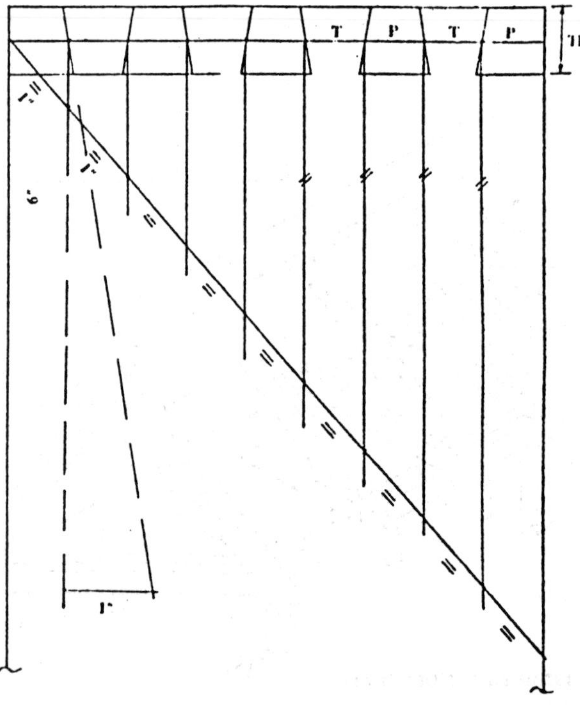

Bay Window Roof Framing

Many people purchase bay windows in standard shapes and sizes that are ready to be installed, but have problems figuring the geometry for the roof framing. This example is based on the typical octagon shaped bay window.

After the window is installed, an additional plate may be necessary to serve as a rafter plat. A plate must also be secured to the wall as a ridge. Angle 15 in Figure 1 is the cut of this plate where it slopes to meet the horizontal. Removing the hip top edge to form a bevel 13-14 is necessary because of the small size of the roof.

This development could be done to scale, but a full size layout will give you all the information you need, including lengths and sheathing shapes. Remember to allow for a rafter projection if it is desired.

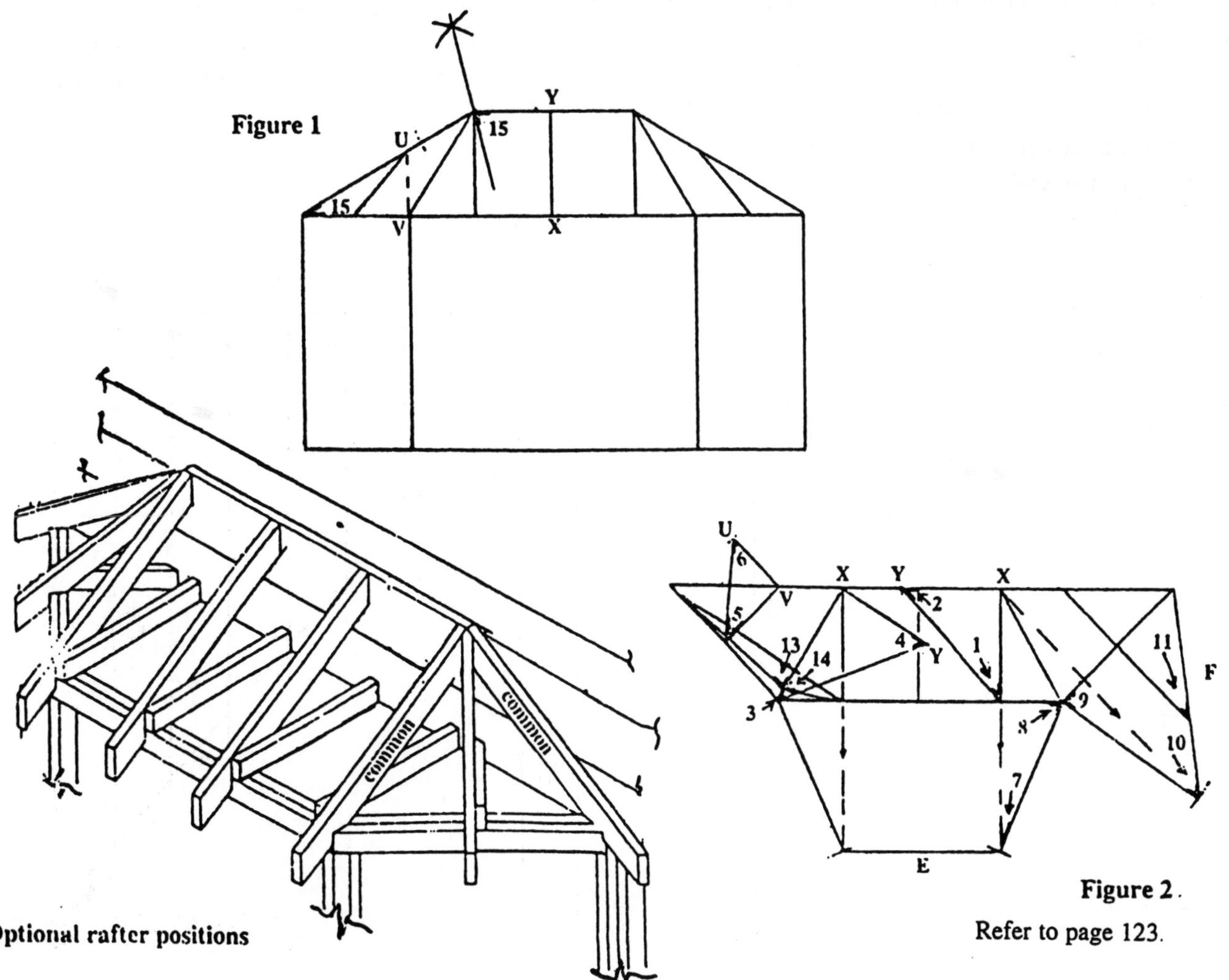

Figure 1

Optional rafter positions

Figure 2.

Refer to page 123.

Development

First draw the plan and elevation.

Next, use roof rise XY and lay out the common rafter on the plan showing its point length, plumb and level cuts at 1-2.

Follow those steps on the plan of hip for 3-4.

For jack rafter length and cuts 5-6, use its rise UV.

Now develop surfaces E and F to give 7-8-9-10 hip side cuts and 11 jack side cuts.

No. 13 and 14 are the dihedral angles (backing bevel) and 15 is the mitre angle for wall plates on the building face.

Unusual Moulding Mitres

Mitres in moulded material offer some interesting ideas. What we may consider to be the most common of mitre joints is of course cut at 45°, such as trim casing around a door or window frame.

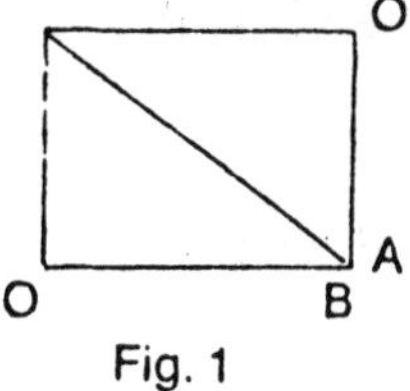

Fig. 1

When two mouldings of different width are mitred, the joint will not be 45°, but the diagonal of a rectangle (Fig. 1) and the sectional proportions will change. In Fig. 2, let A05 be the original profile and B0 the width of the new section. Project the change in section (arrises) of the moulding to the mitre and down to 0B. Next, transfer the original thickness points 0-5 to B0'5' and draw horizontal ordinates to meet the elevation projections. Finally, connect these intersections for the new sectional shape.

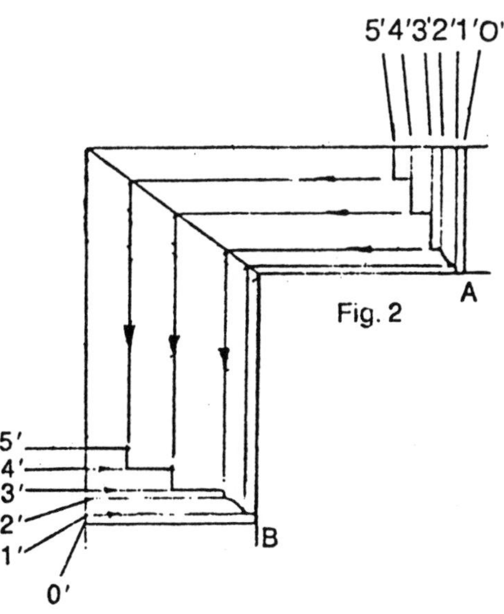

Fig. 2

When dealing with curved mouldings of different sections, the mitre wil be straight: follow above procedure for Fig. 3 (C0 is narrower than B0). However, if the same section is maintained, the mitre will have a slight curve (Fig. 4). To determine this mitre, lay out the curve XY, then set out the same section at A05 and B05. Next, project the elevation (frontal view) arrises of the straight moulding up and beyond the assumed mitre line. Then, using the radial point of the main curve, transfer the arris points of B0. The intersection of these projections will be ordinates to plot the curve.

Note: An extra point was selected midway.

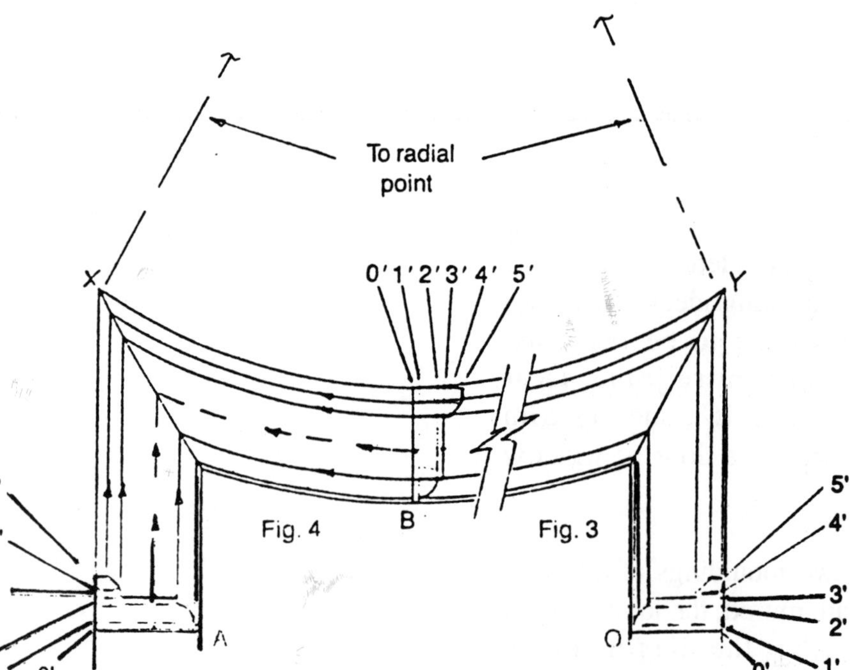

Correct Tooth Angle

The woodworking industry is living in the age of carbide cutters, and this fever for a better cut and a longer life between sharpening has been transmitted to the hobbyist. Most, however, are not prepared to discard their collection of regular saw blades, even if they do own some carbide bits.

For the enthusiast that knows how to touch up (final file) a regular blade, but cannot find the true tooth angle once lost, information like this can prove invaluable.

To determine the correct rake (hook), divide the radius of the blade into six, then use a compass to scribe a circle through the third division. With a straightedge, make a line from a tooth point tangent to the circle; this will give a 30°, rake for the tooth face. The most common angle used for rip-type teeth is 30°, and it is read from a radial line to the centre. By this method the rake can be changed to increase the ripping action, or decreased to create more of a combination cut. Although you may not have the skill to reshape the teeth, at least now you can check for the true tooth angle. If a blade is off more than 3° on the rake or 2° on any other angle, it is not cutting efficiently and should be sent out to be corrected.

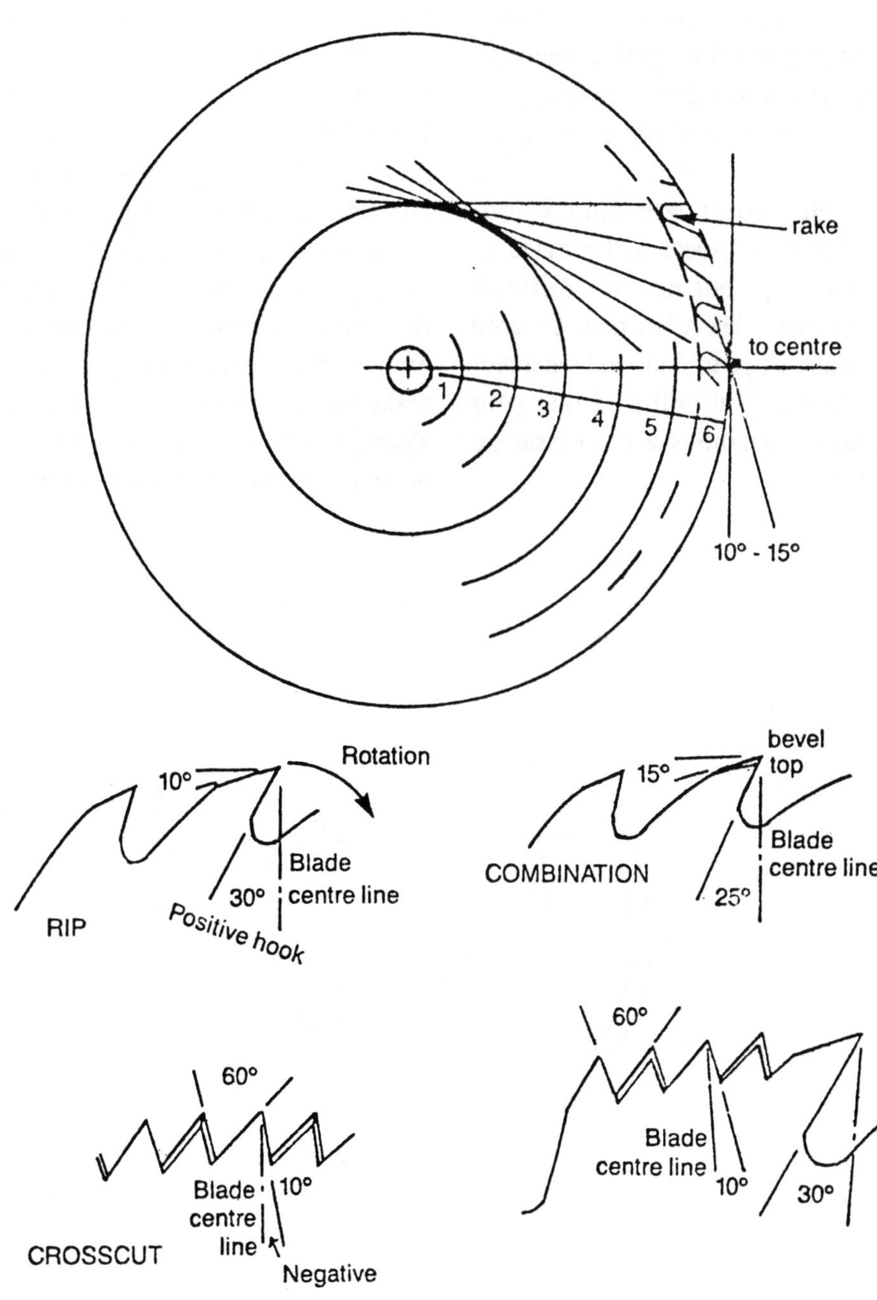

Coffee Table Dish

A geometric design historically used in the era of gothic windows and doorways has equal appeal when incorporated into a table dish.

For the router-wielding woodworker, this project will provide an interesting, but not too difficult challenge. The dish may be used for nuts, chips or fruit and wherever it's in use, you will find that your creation becomes a conversation piece.

Based on foliage (or leaf forms), the individual foils terminate with cusps (a point which returns in itself). Begin by scribing an 18" circle — radius 1. Next, divide the diameter into four equal parts. Then, with the compass point on C, reduce the original radius by 3/4" and scribe the inner circle — radius 2. Transfer the compass to D, the E to make two semicircles — radius 3. These two curves blend with the outer circle and with each other at C. Finally, radius 4 reverses at D and E for the last merging curve. Repeat D and E to complete the other side.

While you're routing, leave a strip through the middle to support the base plate. The can be removed later with a chisel. This dish may be left whole or can be bandsawed across (indicated by arrows) to form two nesting dishes.

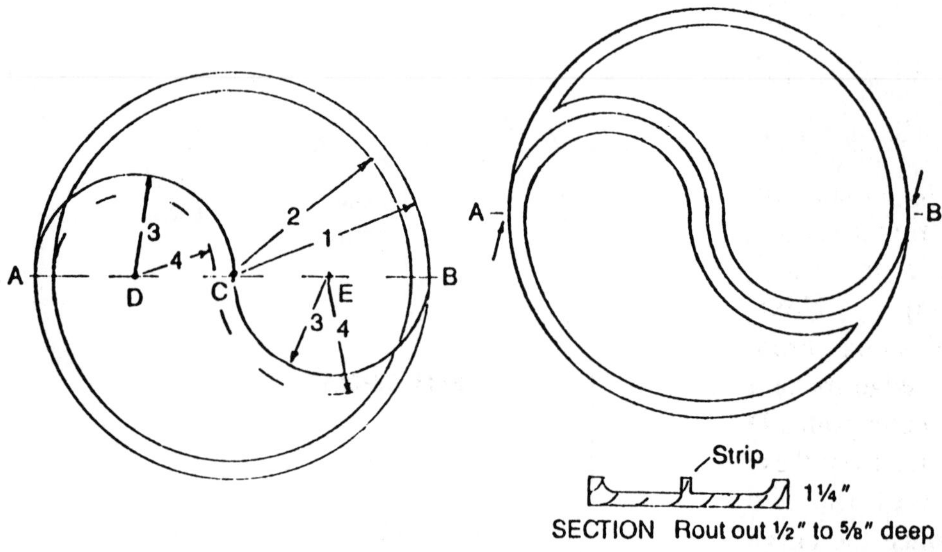

The Garden Trellis

Geometry in the garden is not unusual. The garden trellis is a prime example. Whether it is simple or ornate, its appeal is largely determined by the archway.

Start with two separate ladder frames made from 1-1/2" stock, then design the top bars. Our first example is cut like roof rafters, using 30° for the slope and 60° for the cuts. The second is a half-circle and is cut from three boards per side to keep the board width as narrow as possible. The third idea is cut from 1"x6"x54". To lay out this one, measure 15" from centre and 2-1/2" down to B. Bisect A-B to find centre point C. Now swing an arc with radius A-C, and draw a second arc 2-1/4" inside: let these arcs run random (long). Next, measure a further 9" along. From the top and bottom edge make a 60° angle to form centre points D and E, then scribe arcs from these centres to blend with the two longer curves. Notice the curve from centre E will form the point. Top mounted rails are best for this style because they are more visually pleasing.

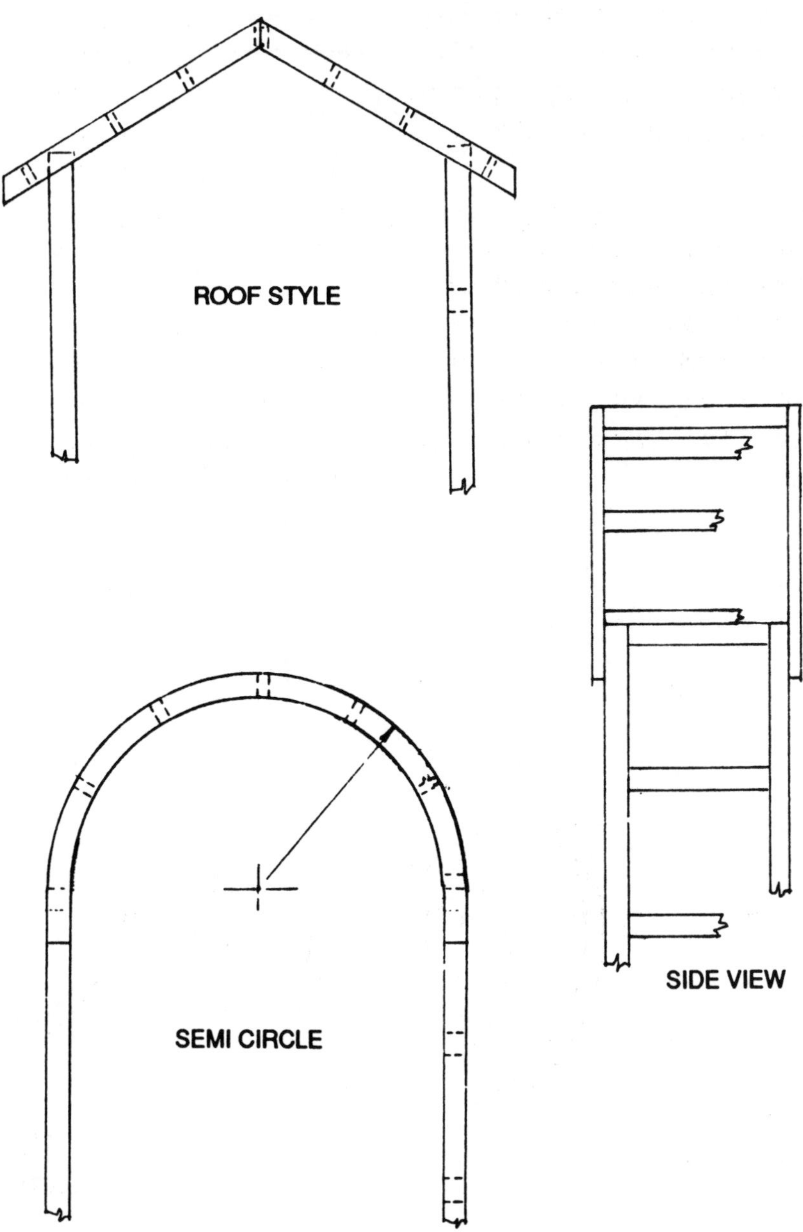

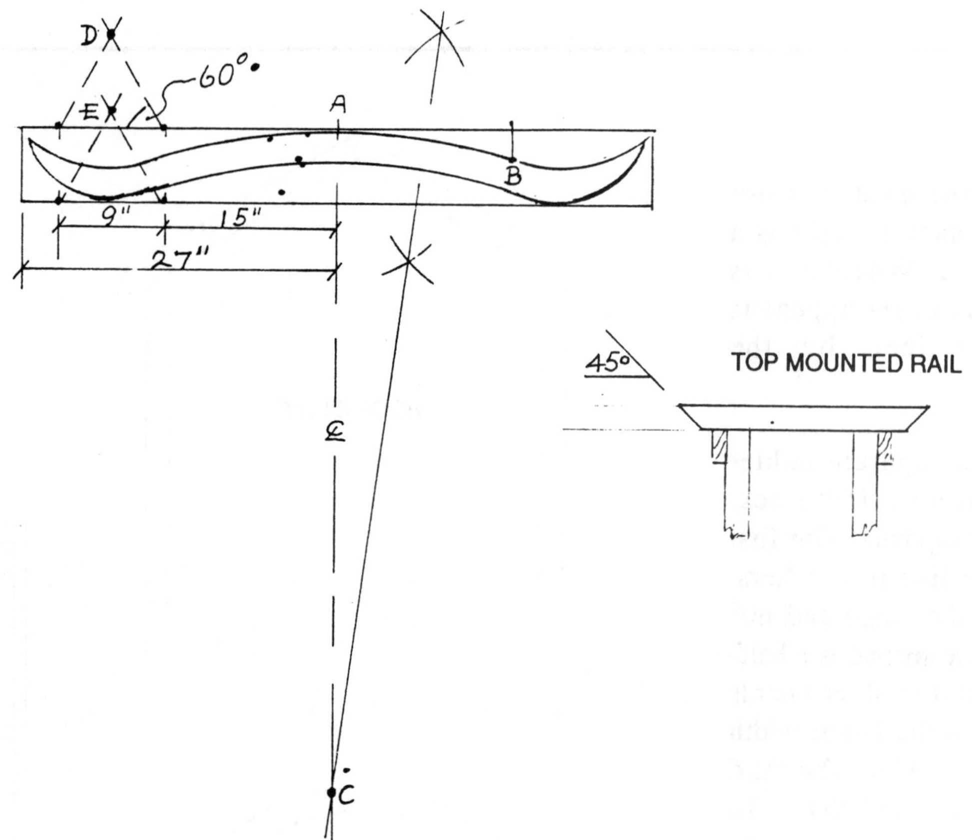
D
E
60°
A
B
9"
15"
27"
℄
C
45°
TOP MOUNTED RAIL

Cupola Construction

The cupola, ornate or plain, is an interesting addition to a roof. Cupolas of various types have had a place in architecture for many years. Though originally intended for ventilation, today they are also used for ornamentation or for extra roof venting. Whatever the reason, they definitely add character to a building. The geometry is quite basic whether built with a frame or constructed from 5/8" to 3/4" plywood.

First, determine the size you would like, keeping in mind that when it is on the roof it may appear smaller than anticipated. (My example measures 30" wide by 30" long with 21" high front and back.) Second, determine the roof slope of the main roof. It is good design to make your cupola roof with the same slope. There are several ways to find this slope: check the original blueprints from which X inches rise per 12 inches of horizontal run; or check the roof with a level and measure down; or measure with a framing square and level at a gable; or simply use an angle protractor. Slope A should be the same as slope A'. The cupola ends will be cut to fit the main roof using the same slope as A".

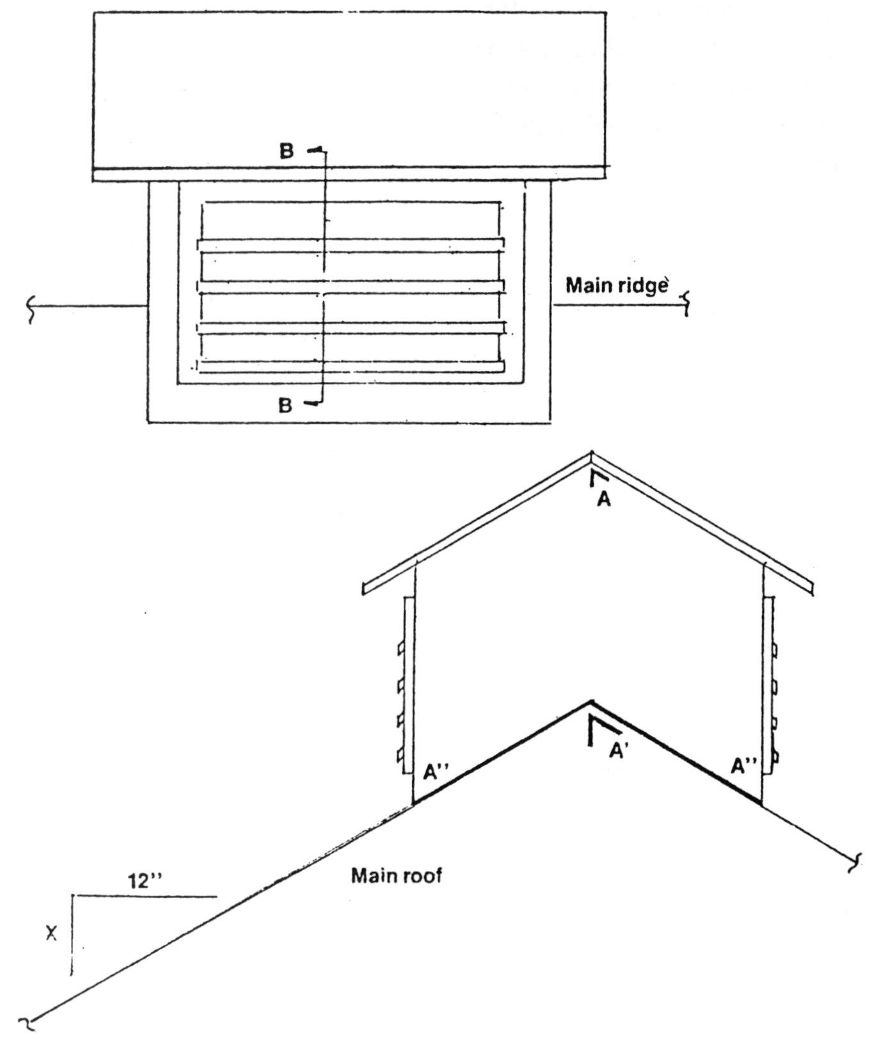

If louvres are required for ventilation (see graphic B-B), then lay out the blade location for dadoes as follows:

1. Decide on size of frame.
2. Lay out top and bottom blade.
3. Decide on number of spaces.
4. Use line division as shown.
5. Lay out remainder of dadoes and cut out. Blades may project to suit. Remember to cover the inside with insect screen.

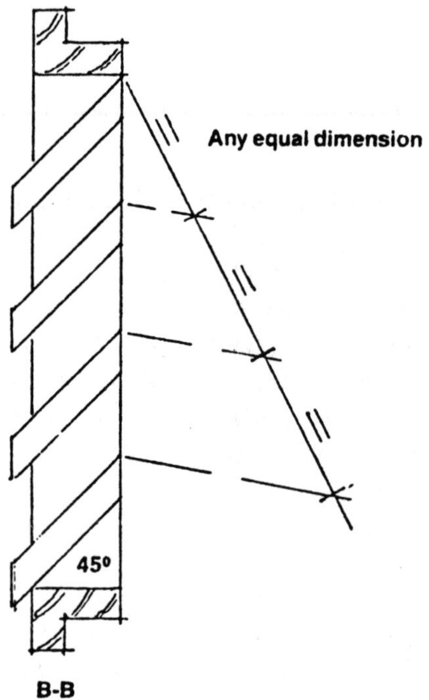

B-B

The Hexagonal Cupola

A hexagonal cupola is a little more involved than the rectangular style. The polygon is based on six 60° triangles. Notice the sides straddle the ridge, rather than put additional strain on wall mitre joints (mitre angle shown at J). Determine suitable dimensions and the main roof slope, then proceed.

Following the principle of surface development for finding true shapes, draw the plan and two elevation views. Use full-size measurements and keep the cupola roof slope the same as the main roof.

From the end view project across (as indicated by the arrows) height H and points of intersection at the roof line. Next, measure three panels using W from the plan, then drop verticals to connect the horizontal projections at E. This gives the true face shape of three sides to fit your original roof slope. Duplicate for opposite sides.

The cupola roof contains six equal triangles developed from the plan drawing. First, pick up the hip length 1-2 from the end view. Transfer this by compass and make arcs from 1' to intersect at 2'. Connect these points for D, the true roof shape. The edge where each panel joins may be left square since it will be covered with shingles.

Fir plywood is a good material for such a project with all joints reinforced as deemed necessary. I suggest decorating the side walls with a moulding, or installing aluminium or wood louvres to vent the main roof.

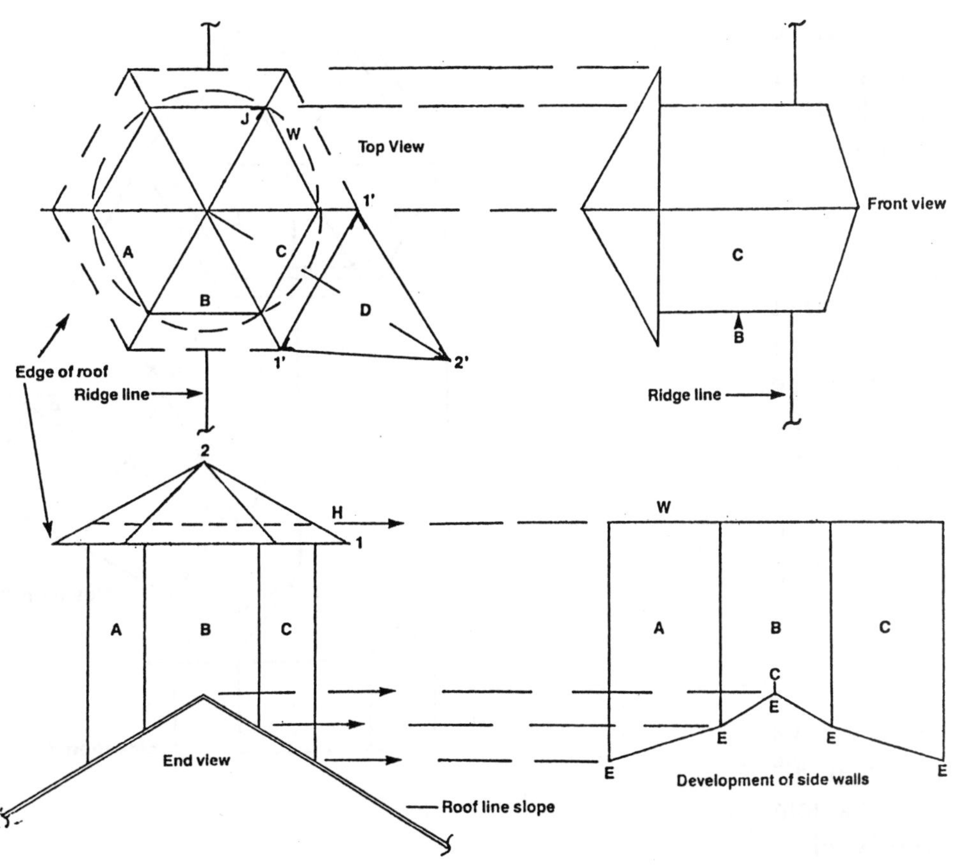

Polygonal Trinket Boxes

Polygons are interesting shapes to work with. Their many sides open up a variety of applications. I have adapted the popular hexagon to a trinket (jewel) box.

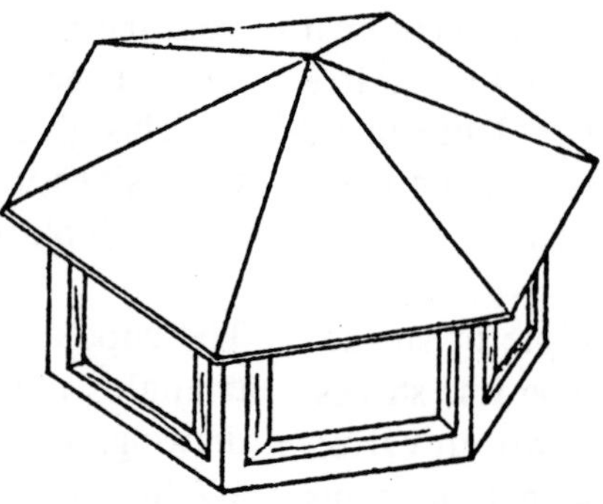

The principle angles required are A, B and C. This polygon is based on six equal triangles of 60° as shown at A, which is also the mitre angle for the side walls of the box. The lid is the real challenge.

For this layout I used dimensions of 5" across the parallel sides of the box by 2-1/2" deep. The lid is 6" from side to side with a 30° slope P.

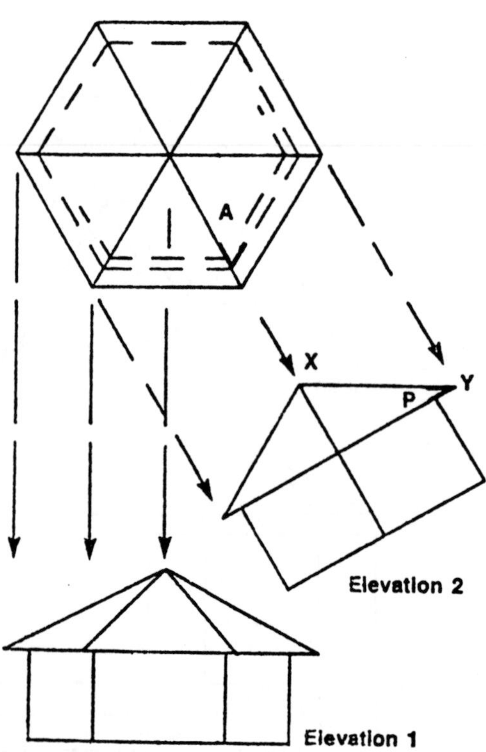

First, lay out the plan and elevation 2 (elevation 1 need not be drawn) using full size measurements. Second, from the edge of the plan at 90° to the centre, measure X-Y and connect B-B to X. This is the true shape for cutting each lid panel. Each of these triangular panels will be mitred to the next, thus requiring an edge bevel C.

To find C, lay out Y-X to give right triangle Y-O-X; then at 90° lay out distance X-O. Now join O to Y giving C the edge bevel.

This joint may be cut on a table saw by setting a mitre gauge to face bevel B and the blade to edge bevel C, thus making a compound bevel in one cut.

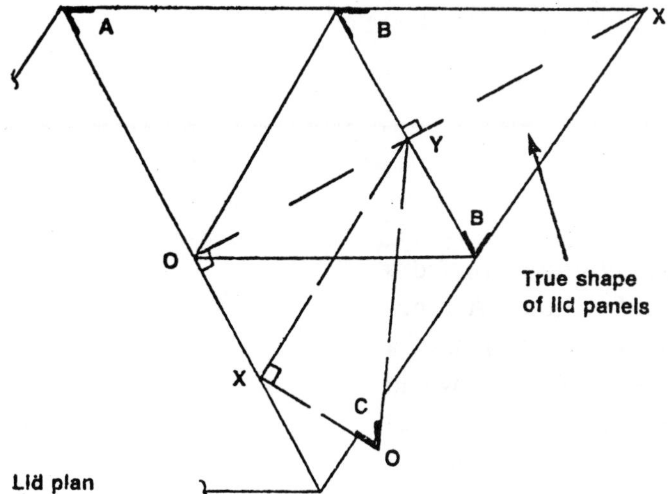

Nonagonal Trinket Boxes

You have already constructed a six-sided trinket box. A nine-sided box, however, is more of a challenge to assemble. (It would probably be worth while to re-read "Polygonal Trinket Boxes" before attempting this one.)

The layout angle X is 40° and the mitre angle 0 is 70°. I used 4" sides to lay out the lid plan and a 30° slope. The walls are inset a 1/2" and are 3" high. The front view need not be drawn. You may choose to follow the sample layout which works for any regular polygon (the sample is a pentagon).

Lay out the lid and use length a-b as the base of a right angle triangle with a hypotenuse at 30° (a"-b"-c). Make another 90° angle at c and mark off c-a" to give e, and triangle c-e-b". Number one is the edge bevel needed to join the triangles together.

The triangular lid shape is found by extending a-b on the plan and marking off the lid length b"-c at b-c". Connect this point back to the corners to find 2, the face bevel.

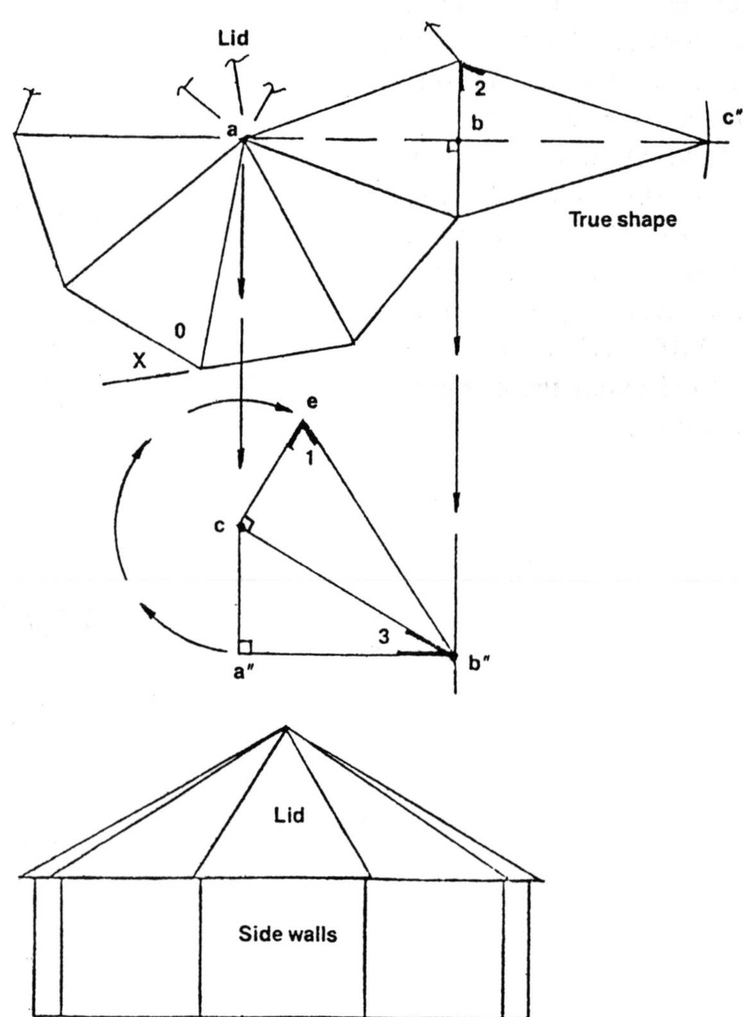

The lower edge of the triangle may be left square or levelled out using angle 3.

To draw any regular polygon on its given side, let AB be the given side of a pentagon. Extend AB to the right. Use B as a centre and use a convenient radius to draw a semi-circle. Divide this into an equal number of parts as the polygon has sides. Through division 2, extend a line BC equal in length to AB (always through 2 to find the second line's inclination). Bisect AB and BC to find 0. Using 0 as centre, scribe a circle to pass through points ABC. AB may now be stepped off about the circle for points D and E.

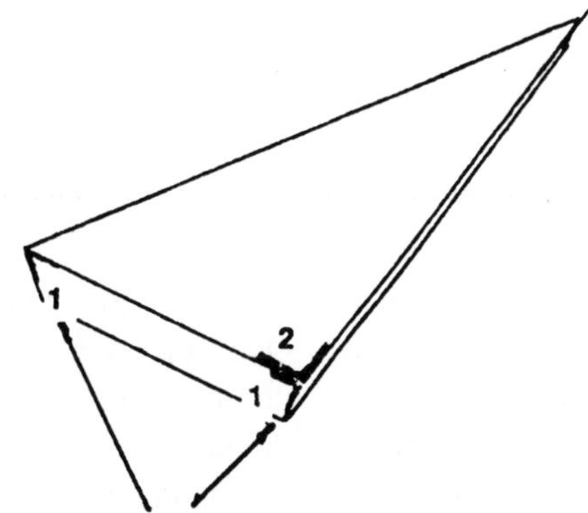

Edge bevel applied

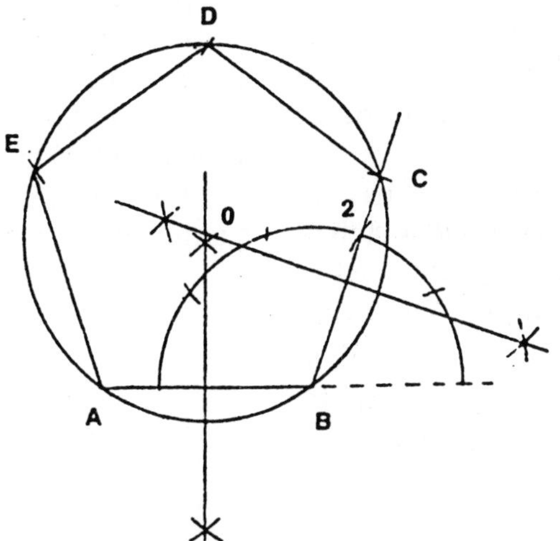

EXAMPLE FOR DRAWING A REGULAR POLYGON ON ITS GIVEN SIDE.

Layout for Concrete Curbs

Concrete curbing is attractive in a cultured garden. Creating the circular curbs for a flower bed requires some geometrical figuring. After deciding on the diameter and curb size, you can proceed with the development if it is required. if you wish to keep the inside and outside surfaces plumb, no development is required, but you will need to calculate the inner and outer circumference.

The example used here has an external diameter of 6' and a curb of 6" x 8" with a slope on the outer face. One quarter of the plan must be laid down along with a cross-section of the curb, both at full size. Following the procedure for cone development, extend the slope to find the apex M of an imaginary cone. Using a slat for a compass and radii M-N and M-O, scribe two arcs. Divide the quarter plan into six or more equal divisions, then step off the same number on the outer arc 0-6. This development N-O and 0-6 gives the pattern for one quarter of the slope which can be used to copy three more. Thin plywood (1/4" or double 1/8") or Masonite will suffice, with stakes every 12" to 16".

Plan B is using straight sides and semi circular ends. In both cases, the beveled face adds more grace to the curb over a vertical one.

Remember, such a curb should be poured on 3" to 4" of gravel in a shallow trench.

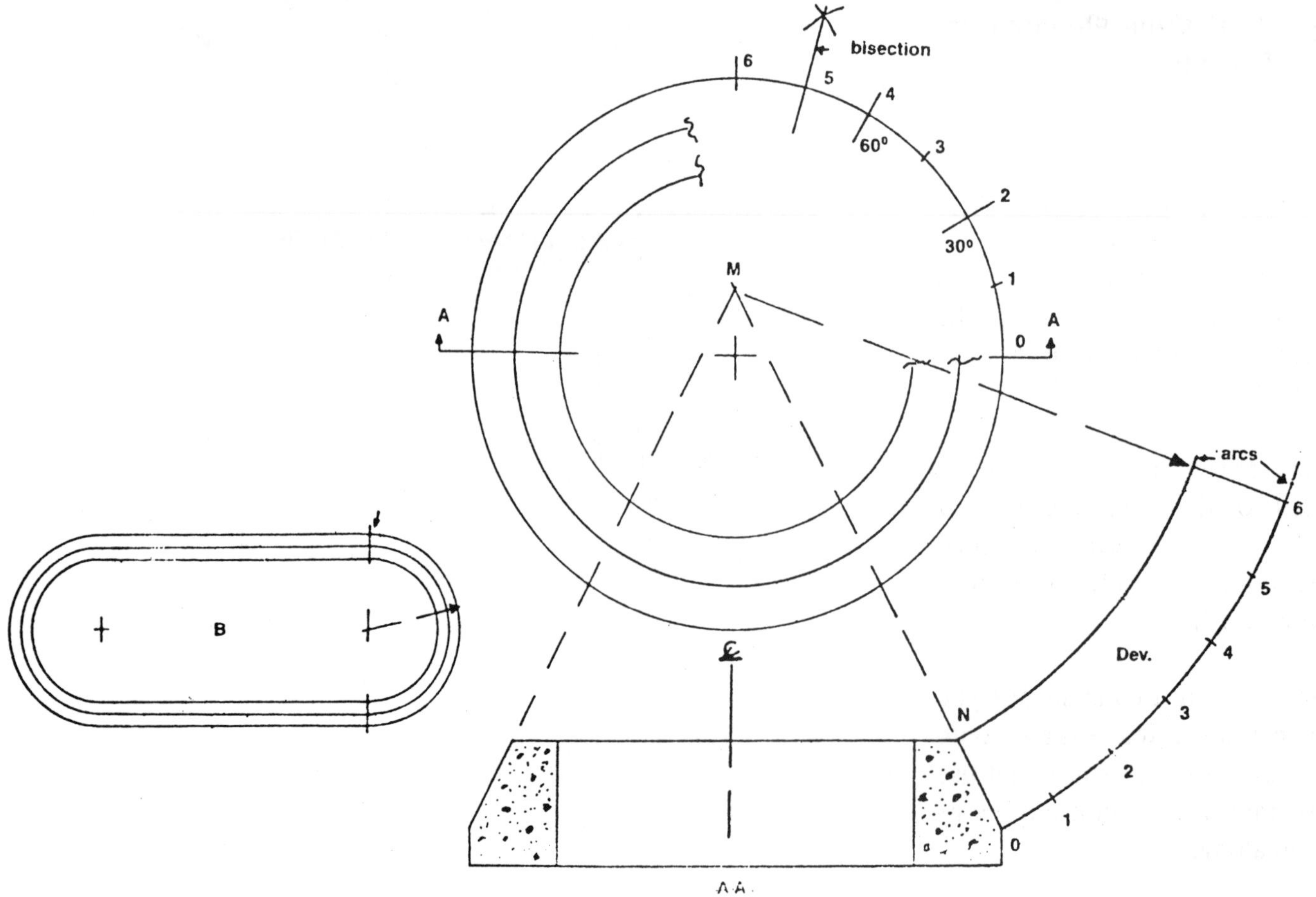

Pattern Transfer

Redrawing free curves from a plain pattern to wood and metal can be a problem. Such curves are called "free" because they have no readily found centres. The problem of getting the right shapes can easily be solved by using the graph method of pattern transfer. This method is not pure applied layout geometry, but it is a practical solution for many jobs.

Here we have two examples —a chair arm and a spindle.

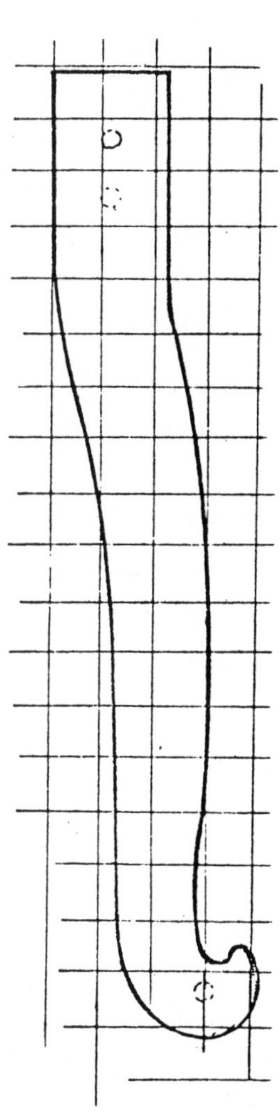

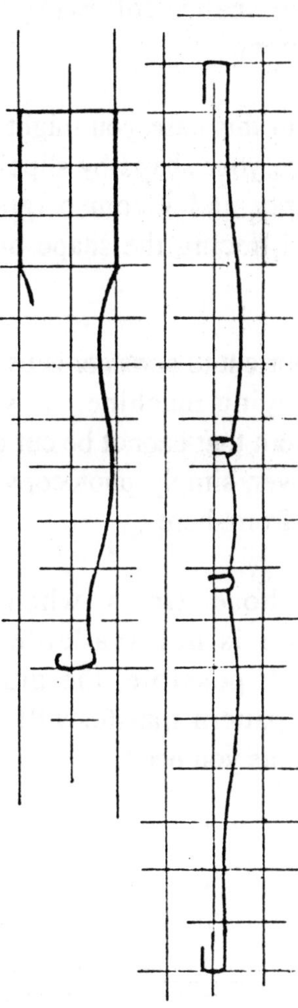

First draw squares on top of the original pattern. In this case I've used 1/2" squares, but the size of the squares depends on the intricacy of the pattern. Choose a size of square that will break the pattern up into manageable segments.

Next draw a grid that is proportional to the size of your new pattern. In this case I wanted a reduction of a half so my new squares are 1/4".

Now simply draw the shapes you find in the original squares onto your new pattern. You might use a protractor to check angles as you go along.

In some cases, you won't want to change the size of a pattern but will want to simply transfer it from a paper pattern to a piece of wood. You follow the same procedure. Draw a grid of appropriate sized squares on your paper pattern, then a similar grid on the wood. Next draw in the shapes following the original pattern.

Of course in this case you might be able to trace your shape by slipping carbon paper under your original pattern and tracing the shape onto the wood.

Another shortcut to consider is using a photocopying machine. If you have a pattern that cannot be cut out or traced over, simply photocopy it, then go on from there.

But for those times when a photocopier is not available or tracing isn't possible, the graph method of pattern transfer will get you the results you need.